TJ
1185
H27
2005

LIBRARY
NSCC, STRAIT AREA CAMPUS
226 REEVES ST.
PORT HAWKESBURY, NS B9A 2A2 CANADA

Advanced Machining Processes

Advanced Machining Processes

Nontraditional and Hybrid Machining Processes

Hassan El-Hofy
Production Engineering Department
Alexandria University, Egypt

McGraw-Hill

New York   Chicago   San Francisco   Lisbon   London   Madrid
Mexico City   Milan   New Delhi   San Juan   Seoul
Singapore   Sydney   Toronto

CIP Data is on file with the Library of Congress.

Copyright © 2005 by The McGraw-Hill Companies, Inc. All rights reserved. Printed in the United States of America. Except as permitted under the United States Copyright Act of 1976, no part of this publication may be reproduced or distributed in any form or by any means, or stored in a data base or retrieval system, without the prior written permission of the publisher.

1 2 3 4 5 6 7 8 9 0 DOC/DOC 0 1 0 9 8 7 6 5

ISBN 0-07-145334-2

The sponsoring editor for this book was Kenneth P. McCombs and the production supervisor was Sherri Souffrance. It was set in Century Schoolbook by International Typesetting and Composition. The art director for the cover was Anthony Landi.

Printed and bound by RR Donnelley.

 This book was printed on recycled, acid-free paper containing a minimum of 50% recycled, de-inked fiber.

McGraw-Hill books are available at special quantity discounts to use as premiums and sales promotions, or for use in corporate training programs. For more information, please write to the Director of Special Sales, McGraw-Hill Professional, Two Penn Plaza, New York, NY 10121-2298. Or contact your local bookstore.

Information contained in this work has been obtained by The McGraw-Hill Companies, Inc. ("McGraw-Hill") from sources believed to be reliable. However, neither McGraw-Hill nor its authors guarantee the accuracy or completeness of any information published herein, and neither McGraw-Hill nor its authors shall be responsible for any errors, omissions, or damages arising out of use of this information. This work is published with the understanding that McGraw-Hill and its authors are supplying information but are not attempting to render engineering or other professional services. If such services are required, the assistance of an appropriate professional should be sought.

Dedicated to my wife Soaad El-Hofy

Contents

Preface   xi
Acknowledgments   xvii
List of Acronyms   xix
List of Symbols   xxiii

Chapter 1. Material Removal Processes   1

- 1.1 Introduction   1
- 1.2 History of Machining   1
- 1.3 Traditional Machining   5
  - 1.3.1 Machining by cutting   5
  - 1.3.2 Machining by abrasion   6
- 1.4 Nontraditional Machining   8
  - 1.4.1 Single-action nontraditional machining   9
  - 1.4.2 Hybrid machining   10
- References   13

Chapter 2. Mechanical Processes   15

- 2.1 Ultrasonic Machining   15
  - 2.1.1 Introduction   15
  - 2.1.2 The machining system   15
  - 2.1.3 Material removal process   22
  - 2.1.4 Factors affecting material removal rate   24
  - 2.1.5 Dimensional accuracy and surface quality   26
  - 2.1.6 Applications   28
- 2.2 Water Jet Machining   32
  - 2.2.1 Introduction   32
  - 2.2.2 The machining system   32
  - 2.2.3 Process parameters   34
  - 2.2.4 Applications   35
  - 2.2.5 Advantages and disadvantages of WJM   38
- 2.3 Abrasive Jet Machining   39
  - 2.3.1 Introduction   39
  - 2.3.2 Machining system   39
  - 2.3.3 Material removal rate   40
  - 2.3.4 Applications   42
  - 2.3.5 Advantages and limitations of AJM   42

2.4	Abrasive Water Jet Machining	43
	2.4.1 Introduction	43
	2.4.2 The machining system	44
	2.4.3 Process capabilities	45
2.5	Ice Jet Machining	46
	2.5.1 Introduction	46
	2.5.2 Process description	46
2.6	Magnetic Abrasive Finishing	48
	2.6.1 Introduction	48
	2.6.2 The machining system	48
	2.6.3 Material removal process	49
	2.6.4 Applications	50
	References	52

Chapter 3. Chemical Processes — 55

3.1	Chemical Milling	55
	3.1.1 Introduction	55
	3.1.2 Tooling for CHM	57
	3.1.3 Process parameters	61
	3.1.4 Material removal rate	61
	3.1.5 Accuracy and surface finish	62
	3.1.6 Advantages	63
	3.1.7 Limitations	64
	3.1.8 Applications	64
3.2	Photochemical Milling	66
	3.2.1 Introduction	66
	3.2.2 Process description	66
	3.2.3 Applications	67
	3.2.4 Advantages	68
3.3	Electropolishing	70
	3.3.1 Introduction	70
	3.3.2 Process parameters	73
	3.3.3 Applications	73
	3.3.4 Process limitations	74
	References	75

Chapter 4. Electrochemical Processes — 77

4.1	Electrochemical Machining	77
	4.1.1 Introduction	77
	4.1.2 Principles of electrolysis	77
	4.1.3 Theory of ECM	78
	4.1.4 ECM equipment	79
	4.1.5 Basic working principles	84
	4.1.6 Process characteristics	87
	4.1.7 Process control	95
	4.1.8 Applications	97
	4.1.9 Micro-ECM	98
	4.1.10 Advantages and disadvantages of ECM	98
	4.1.11 Environmental impacts	99
4.2	Electrochemical Drilling	100
4.3	Shaped Tube Electrolytic Machining	102

4.4	Electrostream (Capillary) Drilling	105
4.5	Electrochemical Jet Drilling	108
4.6	Electrochemical Deburring	109
	References	112

Chapter 5. Thermal Processes — 115

- 5.1 Electrodischarge Machining — 115
  - 5.1.1 Introduction — 115
  - 5.1.2 Mechanism of material removal — 115
  - 5.1.3 The machining system — 120
  - 5.1.4 Material removal rates — 125
  - 5.1.5 Surface integrity — 127
  - 5.1.6 Heat-affected zone — 129
  - 5.1.7 Applications — 130
  - 5.1.8 Process control — 137
  - 5.1.9 EDM automation — 138
  - 5.1.10 Environmental impact — 139
- 5.2 Laser Beam Machining — 140
  - 5.2.1 Introduction — 140
  - 5.2.2 Material removal mechanism — 141
  - 5.2.3 Applications — 144
  - 5.2.4 Advantages and limitations — 156
- 5.3 Electron Beam Machining — 157
  - 5.3.1 Introduction — 157
  - 5.3.2 Basic equipment and removal mechanism — 157
  - 5.3.3 Applications — 163
  - 5.3.4 Advantages and disadvantages — 165
- 5.4 Plasma Beam Machining — 166
  - 5.4.1 Introduction — 166
  - 5.4.2 Machining systems — 166
  - 5.4.3 Material removal rate — 169
  - 5.4.4 Accuracy and surface quality — 169
  - 5.4.5 Applications — 171
  - 5.4.6 Advantages and disadvantages — 172
- 5.5 Ion Beam Machining — 172
  - 5.5.1 Introduction — 172
  - 5.5.2 Material removal rate — 173
  - 5.5.3 Accuracy and surface effects — 175
  - 5.5.4 Applications — 176
- References — 177

Chapter 6. Hybrid Electrochemical Processes — 181

- 6.1 Introduction — 181
- 6.2 Electrochemical Grinding — 182
  - 6.2.1 Introduction — 182
  - 6.2.2 Material removal rate — 183
  - 6.2.3 Accuracy and surface quality — 187
  - 6.2.4 Applications — 188
  - 6.2.5 Advantages and disadvantages — 188
- 6.3 Electrochemical Honing — 189
  - 6.3.1 Introduction — 189
  - 6.3.2 Process characteristics — 189

		6.3.3 Applications	191
	6.4	Electrochemical Superfinishing	192
		6.4.1 Introduction	192
		6.4.2 Material removal process	193
		6.4.3 Process accuracy	195
	6.5	Electrochemical Buffing	196
		6.5.1 Introduction	196
		6.5.2 Material removal process	196
	6.6	Ultrasonic-Assisted ECM	197
		6.6.1 Introduction	197
		6.6.2 Material removal process	198
	6.7	Laser-Assisted ECM	199
		References	201

Chapter 7. Hybrid Thermal Processes — 203

7.1	Introduction	203
7.2	Electroerosion Dissolution Machining	204
7.3	Electrodischarge Grinding	212
7.4	Abrasive Electrodischarge Machining	216
7.5	EDM with Ultrasonic Assistance	218
7.6	Electrochemical Discharge Grinding	221
7.7	Brush Erosion-Dissolution Mechanical Machining	224
	References	226

Chapter 8. Material Addition Processes — 229

8.1	Introduction		229
8.2	Liquid-Based Techniques		230
	8.2.1	Stereolithography	230
	8.2.2	Holographic interference solidification	232
	8.2.3	Beam interference solidification	232
	8.2.4	Solid ground curing	233
	8.2.5	Liquid thermal polymerization	235
	8.2.6	Fused deposition modeling	235
	8.2.7	Multijet modeling	238
	8.2.8	Ballistic particles manufacturing	239
	8.2.9	Shape deposition manufacturing	240
8.3	Powder-Based Processes		241
	8.3.1	Selective laser sintering	241
	8.3.2	Laser engineered net shaping	242
	8.3.3	Three-dimensional printing	243
8.4	Solid-Based Techniques		244
	8.4.1	Solid foil polymerization	244
	8.4.2	Laminated object modeling	245
	References		246

Index 249

Preface

Machining processes produce finished products with a high degree of accuracy and surface quality. Conventional machining utilizes cutting tools that must be harder than the workpiece material. The use of difficult-to-cut materials encouraged efforts that led to the introduction of the nonconventional machining processes that are well-established in modern manufacturing industries.

Single-action nontraditional machining processes are classified on the basis of the machining action causing the material removal from the workpiece. For each process, the material removal mechanism, machining system components, process variables, technological characteristics, and industrial applications are presented.

The need for higher machining productivity, product accuracy, and surface quality led to the combination of two or more machining actions to form a new hybrid machining process. Based on the major mechanism causing the material removal process, two categories of hybrid machining processes are introduced. A review of the existing hybrid machining processes is given together with current trends and research directions. For each hybrid machining process the method of material removal, machining system, process variables, and applications are discussed.

This book provides a comprehensive reference for nontraditional machining processes as well as for the new hybrid machining ones. It is intended to be used for degree and postgraduate courses in production, mechanical, manufacturing, and industrial engineering. It is also useful to engineers working in the field of advanced machining technologies.

In preparing the text, I paid adequate attention to presenting the subject in a simple and easy to understand way. Diagrams are simple and self-explanatory. I express my gratitude to all authors of various books, papers, Internet sites, and other literature which have been referred to in this book. I will be glad to receive comments and suggestions for enhancing the value of this book in future editions.

Outline of the book

The following subjects and chapters are organized as a journey toward understanding the characteristics of nonconventional and hybrid machining processes. The book is written in eight chapters:

Chapter 1: Material Removal Processes
Chapter 2: Mechanical Processes
Chapter 3: Chemical Processes
Chapter 4: Electrochemical Processes
Chapter 5: Thermal Processes
Chapter 6: Hybrid Electrochemical Processes
Chapter 7: Hybrid Thermal Processes
Chapter 8: Material Addition Processes

In Chap. 1, the history and progress of machining is introduced. The difference between traditional and nontraditional machining is explained. Examples for conventional machining by cutting and abrasion are given. Single-action nontraditional machining is classified according to the source of energy causing the material removal process. Hybrid machining occurs as a result of combining two or more machining phases. Hybrid machining is categorized according to the main material removal mechanism occurring during machining.

Chapter 2 covers a wide range of mechanical nontraditional machining processes such as ultrasonic machining (USM), water jet machining (WJM), abrasive water jet machining (AWJM), ice jet machining (IJM), as well as magnetic abrasive finishing (MAF). In these processes the mechanical energy is used to force the abrasives, water jets, and ice jets that cause mechanical abrasion (MA) to the workpiece material.

In Chap. 3, the chemical machining processes such as chemical milling (CHM), photochemical machining (PCM), and electrolytic polishing (EP) are discussed. In these processes the material is mainly removed through chemical dissolution (CD) occurring at certain locations of the workpiece surface.

Chapter 4 deals with electrochemical machining (ECM) and related applications that include electrochemical drilling (ECDR), shaped tube electrolytic machining (STEM), electrostream (ES), electrochemical jet drilling (ECJD), and electrochemical deburring (ECB). The electrochemical dissolution (ECD) controls the rate of material removal.

Machining processes that are based on the thermal machining action are described in Chap. 5. These include electrodischarge machining (EDM), laser beam machining (LBM), electron beam machining (EBM), plasma beam machining (BPM), and ion beam machining (IBM). In most of these processes, material is removed from the workpiece by melting and evaporation. Thermal properties of the machined parts affect the rate of material removal.

Hybrid electrochemical machining processes are dealt with in Chap. 6. Some of these processes are mainly electrochemical with mechanical assistance using mechanical abrasion such as electrochemical grinding (ECG), electrochemical honing (ECH), electrochemical superfinishing (ECS), and electrochemical buffing (ECB). The introduction of ultrasonic assistance enhances the electrochemical dissolution action during ultrasonic-assisted ECM (USMEC). Laser beams activate electrochemical reactions and hence the rate of material removal during laser-assisted electrochemical machining (ECML).

Chapter 7 covers the hybrid thermal machining processes. Electrochemical dissolution (ECD) enhances the electrodischarge erosion action (EDE) during electroerosion dissolution machining (EEDM). Mechanical abrasion encourages the thermal erosion process during electrodischarge grinding (EDG) and abrasive-assisted electrodischarge machining (AEDG and AEDM). Ultrasonic assistance encourages the discharging process during ultrasonic-assisted EDM (EDMUS). Triple-action hybrid machining occurs by combining both electrochemical dissolution (ECD) and mechanical abrasion to the main erosion phase during electrochemical discharge grinding (ECDG).

Material addition processes are covered in Chap. 8. These include a wide range of rapid prototyping techniques that are mainly classified as liquid-, powder-, and solid-based techniques.

Advantages of the book

1. Covers both the nonconventional and hybrid machining processes
2. Classifies the nonconventional machining processes on the basis of the machining phase causing the material removal (mechanical, thermal, chemical, and electrochemical processes)
3. Classifies the hybrid machining processes based on the major mechanism and hence the machining phase causing the material removal from the workpiece into hybrid thermal and hybrid electrochemical processes
4. Presents clearly the principles of material removal mechanisms in nonconventional machining as well as hybrid machining
5. Explains the role of each machining phase (causing the material removal) on the process behavior
6. Describes the machining systems, their main components, and how they work

7. Discusses the role of machining variables on the technological characteristics of each process (removal rate, accuracy, and surface quality)
8. Introduces the material addition processes that use the same principles adopted in material removal by nonconventional processes

This book is intended to help

1. Undergraduates enrolled in production, industrial, manufacturing, and mechanical engineering programs
2. Postgraduates and researchers trying to understand the theories of material removal by the modern machining processes
3. Engineers and high-level technicians working in the area of advanced machining industries

Why did I write the book?

This book presents 28 years of experience including research and teaching of modern machining methods at many universities around the world. My career started early in the academic year 1975–1976 through a senior project related to the effect of some parameters on the oversize of holes produced by ECM. Afterward, I finished my M.S. degree in the field of accuracy of products by electrolytic sinking in the Department of Production Engineering at Alexandria University. As an assistant lecturer I helped to teach about conventional and nonconventional machining.

I spent 4 years on a study leave in the U.K. working toward my Ph.D. at Aberdeen University and 1 year at Edinburgh University. During that time I finished my thesis in the field of hybrid electrochemical arc wire machining (ECAM) under the supervision of Professor J. McGeough. That work was supported by the Wolfson Foundation and the British Technology Group. I had the Overseas Research Student (ORS) award for three successive years which supported me during my research work. Working on a large research team and sharing discussions in regular meetings, I gained more experience related to many advanced and hybrid machining applications such as hybrid ECM-EDM, ECAM drilling, and electrochemical cusp removal. I was a regular steering committee member for the CAPE conference organized by Professor McGeough. I edited two chapters and shared in the writing of chapter 1 of his book *Micromachining of Engineering Materials*.

Throughout my academic career in which I started out as a lecturer and moved up to being a full professor of modern machining processes, I have taught all subjects related to machining in many universities around the world. I have published about 50 research papers related to nonconventional as well as hybrid machining processes. During my work in Qatar University I was responsible for teaching the advanced machining techniques course. Collecting all materials that I had in a book therefore came to my mind. I have been working on this task since the year 2001.

Hassan El-Hofy
Alexandria, Egypt

Acknowledgments

There are many people who have contributed to the development of this book that I cannot name. First of all, I would like to thank Professors H. Youssef and M. H. Ahmed at the University of Alexandria, Egypt, Professors H. Rahmatallh, S. Soliman and O. Saad at the University of Qatar for their support, suggestions and encouragement.

The editorial and production staff at McGraw Hill have my heartfelt gratitude for their efforts in ensuring that the text is accurate and as well designed as possible.

My greatest thanks have to be reserved to my wife Soaad and daughters Noha, Assmaa, and Lina for their support and interest throughout the preparation of the text. Special thanks have to be offered to my son Mohamed for his discussions, suggestions, and the splendid artwork in many parts of the book.

It is with great pleasure that I acknowledge the help of many organizations that gave me permission to reproduce numerous illustrations and photographs in this book:

- Acu-Line Corporation, Seattle, WA
- ASM International, Materials Park, OH
- ASME International, New York, NY
- Extrude Hone, Irwin, PA
- IEE, Stevenage, UK
- Jet Cut Incorporation, Waterloo, ON, Canada
- Jet-Edge, St. Michael, MN
- LCSM-EFPL, Swiss Federal Institute of Technology, Lausanne, Switzerland
- Precision Engineering Journal, Elsevier, Oxford, UK
- TU/e, Eindhoven University of Technology, Netherlands
- Vectron Deburring, Elyria, OH

List of Acronyms

Abbreviation	Description
AEDG	Abrasive electrodischarge grinding
AEDM	Abrasive electrodischarge machining
AFM	Abrasive flow machining
AJM	Abrasive jet machining
AWJM	Abrasive water jet machining
BHN	Brinell hardness number
BIS	Beam interference solidification
BEDMM	Brush erosion dissolution mechanical machining
BPM	Ballistic particles manufacturing
C	Cutting
CAD	Computer-aided design
CAM	Computer-aided manufacturing
CAPP	Computer-assisted process planning
CBN	Cubic boron nitride
CD	Chemical dissolution
CHM	Chemical milling
CIM	Computer-integrated manufacturing
CVD	Carbon vapor deposition
CNC	Computer numerical control
CW	Continuous wave
EBM	Electron beam machining
ECAM	Electrochemical arc machining
ECB	Electrochemical buffing
ECD	Electrochemical dissolution
ECDB	Electrochemical deburring
ECDG	Electrochemical discharge grinding

Abbreviation	Description
ECDM	Electrochemical discharge machining
ECDR	Electrochemical drilling
ECG	Electrochemical grinding
ECH	Electrochemical honing
ECJD	Electrochemical jet drilling
ECM	Electrochemical machining
ECML	Laser-assisted electrochemical machining
ECS	Electrochemical superfinishing
EDE	Electrodischarge erosion
EDG	Electrodischarge grinding
EDM	Electrodischarge machining
EDMUS	Electrodischarge machining with ultrasonic assistance
EDT	Electrodischarge texturing
EEDM	Electroerosion dissolution machining
EP	Electropolishing
ES	Electrostream
FDM	Fused deposition modeling
FJ	Fluid jet
G	Grinding
HAZ	Heat-affected zone
HF	Hone forming
HIS	Holographic interference solidification
IBM	Ion beam machining
IJM	Ice jet machining
LAE	Laser-assisted chemical etching
LAJECM	Laser-assisted jet ECM
LAN	Local area network
LBM	Laser beam machining
LBT	Laser beam texturing
LENS	Laser engineered net shaping
LOM	Laminated object modeling
LTP	Liquid thermal polymerization
MA	Mechanical abrasion
MAF	Magnetic abrasive finishing
MJM	Multijet modeling
MMC	Metal matrix composites
MPEDM	Mechanical pulse electrodischarge machining

Abbreviation	Description
MRR	Material removal rate
MS	Mechanical scrubbing
MUSM	Micro-ultrasonic machining
NC	Numerical control
ND-YAG	Neodymium-doped yitrium-aluminum-garnet
PAM	Plasma arc machining
PBM	Plasma beam machining
PCB	Photochemical blanking
PCD	Polycrystalline diamond
PECM	Pulse electrochemical machining
PF	Photoforming
PCM	Photochemical milling
PM	Pulsed mode
RP	Rapid prototyping
RUM	Rotary ultrasonic machining
SB	Shot blasting
SDM	Shape deposition manufacturing
SFF	Solid free-form fabrication
SFP	Solid foil polymerization
SGC	Solid ground curing
SLA	Stereolithography
SLS	Selective laser sintering
STEM	Shaped tube electrolytic machining
TEM	Thermal energy method
US	Ultrasonic
USM	Ultrasonic machining
USMEC	Ultrasonic-assisted electrochemical machining
VRR	Volumetric removal rate
WJM	Water jet machining

List of Symbols

Symbol	Definition	Unit
a	tool feed rate	mm/min
A	atomic weight	
A/Z	chemical equivalent	g
$A/Z.F$	electrochemical equivalent	g/C
A_b	area of laser beam at focal point	mm^2
C	electrochemical machining constant	mm^2/s
C/y_e	metal removal rate per unit area	mm/min
C_d	diametrical overcut	mm
C_l	constant depending on material and conversion efficiency	
C_s	speed of sound in magnetostrictor material	m/s
d	CHM undercut	mm
D	EDM depth	mm
D/L_c	corner wear ratio	mm
D/L_e	end wear ratio	
D/L_s	side wear ratio	
d/T	etch factor	
d_a	mean diameter of abrasive particles	μm
d_b	beam diameter at contact with workpiece (slot width)	mm
d_s	spot size diameter	mm
d_t	tool diameter	mm
d_w	produced hole diameter	mm
dy/dt	workpiece rate of change of position	mm/min
E	Young's modulus	MPa
ϵ_m	coefficient of magnetostriction elongation	

List of Symbols

Symbol	Definition	Unit
E_m	magnitudes of magnetic energy	
e_v	number of pulses	
E_v	vaporization energy of material	W/mm^3
E_w	magnitudes of mechanical energy	
$\mathcal{F}$	frequency of oscillation	Hz
f	frequency of changes in magnetic field	Hz
F	Faraday's constant	C/g per-ion
F_l	focal length of lens	cm
f_p	frequency of pulses	Hz
f_r	resonance frequency	Hz
g	depth of hole required	
g_e	depth of hole removed per pulse	mm
g_w	wheel-workpiece gap	
h	thickness of material	mm
H	magnetic field intensity	
H_0	surface fracture strength	BHN
H_{rms}	surface roughness	μm
H_w	hardness of workpiece material	N/mm^2
i	EDM current	A
I	electrolyzing current	A
I_e	beam emission current	mA
I_p	pulse current	A
J	current density	A/mm^{-2}
K	constant	
K_h	constant	μm/μJ
K_m	coefficient of magnetomechanical coupling	
K_p	coefficient of loss	
l	original length of magnetostrictor	
L	slot length	mm
L_c	corner wear	mm
L_e	end wear	mm
L_p	laser power	W
L_s	side wear	mm
m	amount of mass dissolved	g
n	density of target material	atoms per cm^3
n_e	number of pulses	
N	number of abrasives impacting per unit area	

Symbol	Definition	Unit
N_M	relative machinability	
P	density of magnetostrictor material	kg/m^3
P_d	power density	W/cm^2
P_r	pulse power	W
q_c	specific removal rate for pure metals	mm^3/(min·A)
Q_{ECD}	removal rate of electrochemical dissolution	mm^3/min
Q_{ECG}	total removal rate in ECG	mm^3/min
Q_l	linear removal rate	mm/min
Q_{MA}	removal rate of mechanical abrasion	mm^3/min
Q_v	volumetric removal rate	mm^3/min
R	mean radius of grit	mm
R_a	average roughness	μm
R_t	maximum peak to valley roughness	μm
R_w	wear ratio	
S	static stress on tool	kg/mm^2
$s(\theta)$	IBM yield	atoms per ion
t	machining time	min
T	CHM depth of cut	mm
t_i	pulse interval	μs
t_m	machining time	
t_p	pulse duration	μs
T_r	ratio of workpiece to tool electrode melting points	
T_t	melting point of tool electrode	°C
T_w	melting point of workpiece material	°C
U	mean velocity	
V	gap voltage	V
$V(\theta)$	etch rate	atoms per min/mA cm^2
V_a	beam accelerating voltage	kV
V_e	volume of electrode consumed	mm^3
V_g	grinding wheel penetration speed	mm^3/min
VRR	material removal rate	mm^3/min
V_s	machining rate	mm^2/min
V_w	volume of workpiece removed	mm^3
V_w / V_e	volume wear ratio	
W	pulse energy	μJ

Symbol	Definition	Unit
W_t	wear rate of tool	mm³/min
x, y, z	workpiece coordinates	mm
y	gap length	mm
Y	amplitude of vibration	mm
y_e	equilibrium gap	mm
Z	workpiece valence	

Greek symbols

Symbol	Definition	Unit
α	beam divergence	rad
γ	current efficiency of dissolution process	%
Δl	incremental length of magnetostrictor	
ΔT	pulse duration of laser	s
Δv	polarization voltage	V
ε	chemical equivalent weight	
κ	electrolyte conductivity	$\Omega^{-1} \cdot mm^{-1}$
λ	wavelength	
ν	velocity of abrasive particles	m/s
ρ	density of anode material	g/mm³
ρ_a	density of abrasive particles	g/mm³
ρ_e	density of electrolyte	
Ψ	drilling rate	

Advanced Machining Processes

Chapter 1

Material Removal Processes

1.1 Introduction

Parts manufactured by casting, forming, and various shaping processes often require further operations before they are ready for use or assembly. In many engineering applications, parts have to be interchangeable in order to function properly and reliably during their expected service lives; thus control of the dimensional accuracy and surface finish of the parts is required during manufacture. Machining involves the removal of some material from the workpiece (machining allowance) in order to produce a specific geometry at a definite degree of accuracy and surface quality.

1.2 History of Machining

From the earliest of times methods of cutting materials have been adopted using hand tools made from bone, stick, or stone. Later, hand tools made of elementary metals such as bronze and iron were employed over a period of almost one million years. Indeed up to the seventeenth century, tools continued to be either hand operated or mechanically driven by very elementary methods. By such methods, wagons, ships, and furniture, as well as the basic utensils for everyday use, were manufactured. The introduction of water, steam, and, later, electricity as useful sources of energy led to the production of power-driven machine tools which rapidly replaced manually driven tools in many applications.

Based on these advances and together with the metallurgical development of alloy steels as cutting tool materials, a new machine tool industry began to arise in the eighteenth and nineteenth centuries. A major original contribution to this new industry came from John Wilkinson in 1774. He constructed a precision machine for boring engine cylinders, thereby overcoming a problem associated with the first machine tools, which were powered by steam. Twenty-three years later, Henry Maudslay made a further advancement in machining when he devised a screw-cutting engine lathe. James Nasmyth invented the second basic machine tool for shaping and planing; these techniques are used to machine flat surfaces, grooves, shoulders, T-slots, and angular surfaces using single-point cutting tools. The familiar drilling machine is the third category of machine tools; it cuts holes with a twist drill.

Whitney in about 1818 introduced the first milling machine to cut grooves, dovetails, and T-slots as well as flat surfaces. The first universal milling machine, constructed in 1862 by J. R. Brown, was employed to cut helical flutes of twist drills. In the late nineteenth century, the grinding machine was introduced. An advanced form of this technology is the lapping process used to produce a high-quality surface finish and a very tight tolerance, as small as ± 0.00005 millimeters (mm) compared to the ± 0.0025 mm achieved during grinding. Band saws and circular discsaws are used for cutting shapes in metal plates, for making external and internal contours, and for making angular cuts.

A notable development includes the turret lathe made in the middle of the nineteenth century for the automatic production of screws. Another significant advance came in 1896, when F. W. Fellows built a machine that could produce any kind of gear. An example of the significance of early achievements in grinding technology came from C. N. Norton's work in reducing the time needed to grind a car crankshaft from 5 hours (h) to 15 minutes (min). Multiple-station vertical lathes, gang drills, production millers, and special-purpose machines (for example, for broaching, honing, and boring) are other noteworthy examples of advances in machine tool technology (McGeough, 1988). In the later part of the nineteenth century and in the twentieth century, machine tools became increasingly powered by electricity rather than steam. The basic machine tools underwent further refinement; for instance, multiple-point cutters for milling machines were introduced. Even with these advances, conventional machine tool practice still relies on the principle whereby the tool must be made of a material that is harder than the workpiece that is to be cut.

During machining by these conventional methods the operator is given a drawing of the finished part. He or she determines the machining strategy, sets up the machine, and selects tooling, speeds, and feeds. The operator manipulates the machine control to cut the part that passes inspection. Under such circumstances, the product accuracy and surface quality are not satisfactory. Further developments for these conventional machines came by the introduction of copying techniques, cams, and automatic mechanisms that reduced labor and, consequently, raised the product accuracy.

The introduction of numerical control (NC) technology in 1953 opened wide doors to computer numerical control (CNC) and direct numerical control (DNC) machining centers that enhanced the product accuracy and uniformity. Developments in machining processes and their machine tools have continued throughout the last 50 years due to the rapid enhancements in the electronics and computer industries. Ingenious designs of conventional machine tools have enabled complex shapes to be produced at an accuracy of ±1 micrometers (μm). As shown in Fig. 1.1, the most recent developments in conventional machining include precision jig borers, jig grinding, and superfinishing machines. These made the accuracy level of ±1 μm possible. Such a high level of accuracy can be measured using pneumatic or electronic instruments as well as optical comparators. Future trends may also include precision grinding and lapping machines as well as precision diamond lathes.

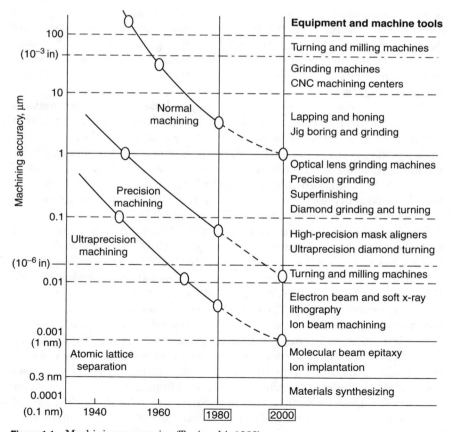

Figure 1.1 Machining accuracies (Tanigushi, 1983).

In modern machining practice, harder, stronger, and tougher materials that are more difficult to cut are frequently used. More attention is, therefore, directed toward machining processes where the mechanical properties of the workpiece material are not imposing any limits on the material removal process. In this regard, the nonconventional machining techniques came into practice as a possible alternative concerning machinability, shape complexity, surface integrity, and miniaturization requirements. Innovative machining techniques or modifications to the existing method by combining different machining processes were needed. Hybrid machining made use of the combined or mutually enhanced advantages and avoided the adverse effects of the constituent processes produced when they are individually applied.

For a while, there were trends toward reducing the workpiece size and dimensions after it became possible to drill ultrasmall-diameter holes (10–100 μm) in hard materials using the available machining processes. Micromachining has recently become, an important issue for further reduction of workpiece size and dimensions. It refers to the technology and practice of making three-dimensional shapes, structures, and devices with dimensions on the order of micrometers. One of the main goals of the development of micromachining is to integrate microelectronics circuitry into micromachined structures and produce completely integrated systems.

Recent applications of micromachining include silicon micromachining, excimer lasers, and photolithography. Machines such as precision grinders may be capable of producing an accuracy level of ±0.01 μm that can also be measured using laser instruments, and optical fibers. Future trends in micromachining include laser and electron beam lithography and superhigh-precision grinding, lapping, and polishing machines. In such cases high-precision laser beam measuring instruments are used as indicated by McGeough (2002).

The desired high-precision nanomachining requirements can be obtained by removing atoms or molecules rather than chips as in the case of ion beam machining. Nanomachining was introduced by Tanigushi (1983) to cover the miniaturization of components and tolerances in the range from the submicron level down to that of an individual atom or molecule between 100 nanometers (nm) and 0.1 nm. The need for such a small scale arose for the high performance and efficiency required in many fields such as microelectronics and in the automobile and aircraft manufacturing industries. The achievable accuracy of nanomachining has increased by almost two orders of magnitude in the last decade. Nanomachining processes include atom, molecule, or ion beam machining, and atom or molecule deposition. These techniques can achieve ±1-nm tolerances that can be measured using a scanning electron microscope (SEM), a transmission electron microscope, an ion analyzer, or electron diffraction equipment.

1.3 Traditional Machining

As mentioned earlier, machining removes certain parts of the workpieces to change them to final parts. Traditional, also termed conventional, machining requires the presence of a tool that is harder than the workpiece to be machined. This tool should be penetrated in the workpiece to a certain depth. Moreover, a relative motion between the tool and workpiece is responsible for forming or generating the required shape. The absence of any of these elements in any machining process such as the absence of tool-workpiece contact or relative motion, makes the process a nontraditional one. Traditional machining can be classified according to the machining action of cutting (C) and mechanical abrasion (MA) as shown in Fig. 1.2.

1.3.1 Machining by cutting

During machining by cutting, the tool is penetrated in the work material to the depth of the cut. A relative (main and feed) motion determines the workpiece geometry required. In this regard, turning produces cylindrical parts, shaping and milling generate flat surfaces, while drilling

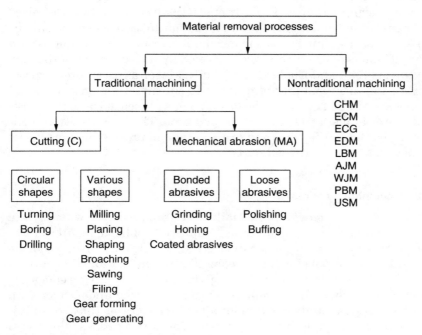

Figure 1.2 Material removal processes.

produces holes of different diameters. Tools have a specific number of cutting edges of a known geometry. The cutting action removes the machining allowance in the form of chips, which are visible to the naked eye. During machining by cutting, the shape of the workpiece may be produced by forming when the cutting tool possesses the finished contour of the workpiece. A relative motion is required to produce the chip (main motion) in addition to the tool feed in depth as shown in Fig. 1.3a. The accuracy of the surface profile depends mostly on the accuracy of the form-cutting tool. A surface may also be generated by several motions that accomplish the chip formation process (main motion) and the movement of the point of engagement along the surface (feed motion). Fig. 1.3b provides a typical example of surface generation by cutting. Slot milling, shown in Fig. 1.3c, adopts the combined form and generation cutting principles.

The resistance of the workpiece material to machining by cutting depends on the temperature generated at the machining zone. High-speed hot machining is now recognized as one of the key manufacturing techniques with high productivity. As the temperature rises, the strength decreases while the ductility increases. It is quite logical to assume that the high temperature reduces the cutting forces and energy consumption and enhances the machinability of the cut material. Hot machining has been employed to improve the machinability of glass and engineering ceramics. El-Kady et al. (1998) claimed that workpiece heating is intended not only to reduce the hardness of the material but also to change the chip formation mechanism from a discontinuous chip to a continuous one, which is accompanied by improvement of the surface finish. Todd and Copley (1997) built a laser-assisted prototype to improve the machinability of difficult-to-cut materials on traditional turning and milling centers. The laser beam was focused onto the workpiece material just above the machining zone. The laser-assisted turning reduced the cutting force and tool wear and improved the geometrical characteristics of the turned parts.

1.3.2 Machining by abrasion

The term *abrasion machining* usually describes processes whereby the machining allowance is removed by a multitude of hard, angular abrasive particles or grains (also called grits), which may or may not be bonded to form a tool of definite geometry. In contrast to metal cutting processes, during abrasive machining, the individual cutting edges are randomly oriented and the depth of engagement (the undeformed chip thickness) is small and not equal for all abrasive grains that are simultaneously in contact with the workpiece. The cutting edges (abrasives) are used to remove a small machining allowance by the MA action

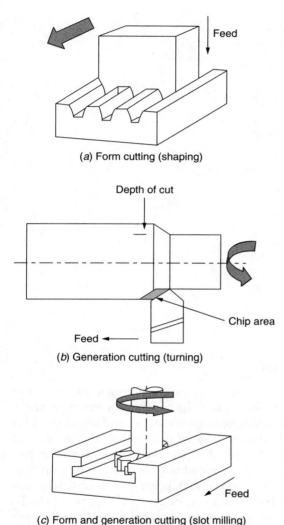

(a) Form cutting (shaping)

(b) Generation cutting (turning)

(c) Form and generation cutting (slot milling)

Figure 1.3 Metal cutting processes.

during the finishing processes. The material is removed in the form of minute chips, which are invisible in most cases (Kaczmarek, 1976). The MA action is adopted during grinding, honing, and superfinishing processes that employ either solid grinding wheels or sticks in the form of bonded abrasives (Fig. 1.4a). Furthermore, in lapping, polishing, and buffing, loose abrasives are used as tools in a liquid machining media as shown in Fig. 1.4b.

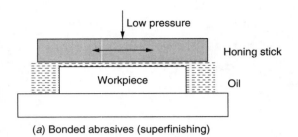

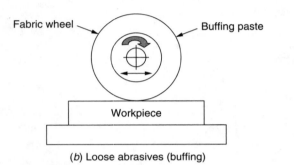

Figure 1.4 Abrasive machining.

1.4 Nontraditional Machining

The greatly improved thermal, chemical, and mechanical properties of the new engineering materials made it impossible to machine them using the traditional machining processes of cutting and abrasion. This is because traditional machining is most often based on the removal of material using tools that are harder than the workpiece. For example, the high ratio of the volume of grinding wheel worn per unit volume of metal removed (50–200) made classical grinding suitable only to a limited extent for production of polycrystalline diamond (PCD) profile tools.

The high cost of machining ceramics and composites and the damage generated during machining are major obstacles to the implementation of these materials. In addition to the advanced materials, more complex shapes, low-rigidity structures, and micromachined components with tight tolerances and fine surface quality are often needed. Traditional machining methods are often ineffective in machining these parts. To meet these demands, new processes are developed.

These methods play a considerable role in the aircraft, automobile, tool, die, and mold making industries. The nontraditional machining methods (Fig. 1.5) are classified according to the number of machining actions causing the removal of material from the workpiece.

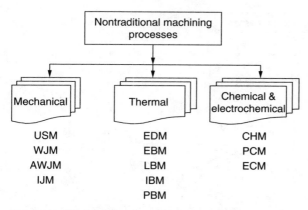

Figure 1.5 Nontraditional machining processes.

1.4.1 Single-action nontraditional machining

For these processes only one machining action is used for material removal. These can be classified according to the source of energy used to generate such a machining action: mechanical, thermal, chemical, and electrochemical.

1.4.1.1 Mechanical machining. Ultrasonic machining (USM) and water jet machining (WJM) are typical examples of single-action, mechanical, nontraditional machining processes. Machining occurs by MA in USM while cutting is adopted using a fluid jet in case of WJM. The machining medium is solid grains suspended in the abrasive slurry in the former, while a fluid is employed in the WJM process. The introduction of abrasives to the fluid jet enhances the cutting in case of abrasive water jet machining (AWJM) or ice particles during ice jet machining (IJM) (see Fig. 1.6).

1.4.1.2 Thermal machining. Thermal machining removes the machining allowance by melting or vaporizing the workpiece material. Many secondary phenomena relating to surface quality occur during machining such as microcracking, formation of heat-affected zones, and striations.

The source of heat required for material removal can be the plasma during electrodischarge machining (EDM) and plasma beam machining (PBM), photons during laser beam machining (LBM), electrons in case of electron beam machining (EBM), or ions for ion beam machining (IBM). For each of these processes, the machining medium is different.

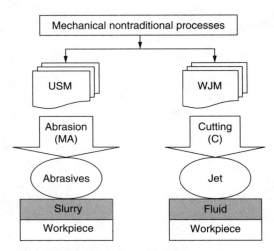

Figure 1.6 Mechanical machining processes.

While electrodischarge occurs in a dielectric liquid for EDM, ion and laser beams are achieved in a vacuum during IBM and LBM as shown in Fig. 1.7.

1.4.1.3 Chemical and electrochemical machining. Chemical milling (CHM) and photochemical machining (PCM), also called chemical blanking (PCB), use a chemical dissolution (CD) action to remove the machining allowance through ions in an etchant. Electrochemical machining (ECM) uses the electrochemical dissolution (ECD) phase to remove the machining allowance using ion transfer in an electrolytic cell (Fig. 1.8).

1.4.2 Hybrid machining

Technological improvement of machining processes can be achieved by combining different machining actions or phases to be used on the material being removed. A mechanical conventional single cutting or MA action process can be combined with the respective machining phases of electrodischarge (ED) in electrodischarge machining (EDM) or ECD in ECM. The reason for such a combination and the development of a hybrid machining process is mainly to make use of the combined advantages and to avoid or reduce some adverse effects the constituent processes produce when they are individually applied. The performance characteristics of a hybrid process are considerably different from those of the single-phase processes in terms of productivity, accuracy, and surface quality (*www.unl.edu.nmrc/outline.htm*).

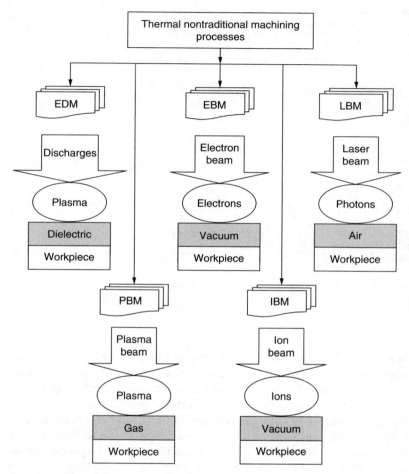

Figure 1.7 Thermal nonconventional processes.

Depending on the major machining phase involved in the material removal, hybrid machining can be classified into hybrid chemical and electrochemical processes and hybrid thermal machining.

1.4.2.1 Hybrid chemical and electrochemical processes. In this family of hybrid machining processes, the major material removal phase is either CD or ECD. Such a machining action can be combined with the thermal assistance by local heating in case of laser-assisted electrochemical machining (ECML). In other words, the introduction of the mechanical abrasion action assists the ECD machining phase during electrochemical grinding (ECG) and electrochemical superfinishing (ECS).

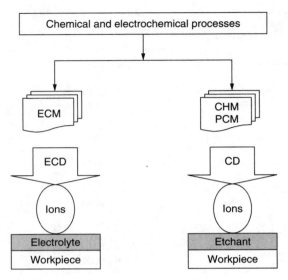

Figure 1.8 Electrochemical and chemical machining processes.

Ultrasonic-assisted electrochemical machining (USMEC) employs an USM component with ECM. The mechanical action of the fluid jet assists the process of chemical dissolution in electrochemical buffing (ECB).

Kozak and Rajurkar (2000) reported that the mechanical interaction with workpiece material changes the conditions for a better anodic dissolution process through mechanical depassivation of the surface. Under such conditions, removing thin layers of oxides and other compounds from the anode surface makes the dissolution and smoothing processes more intensive. Significant effects of the mechanical machining action have been observed with ultrasonic waves. The cavitations generated by such vibrations enhance the ECM by improving electrolyte flushing and hence the material removal from the machined surface.

1.4.2.2 Hybrid thermal machining. In this case the main material removal mechanism is a thermal one. The combination of this phase with the ECD phase, MA action, and ultrasonic (US) vibration generates a family of double action processes. The triplex hybrid machining is also achievable by combining the electrodischarge erosion (EDE) phase, the ECD action, and the MA in grinding (G). Such a combination enhance the rate of material removal and surface quality in electrochemical discharge grinding (ECDG) and the other hybrid processes shown in Fig. 1.9.

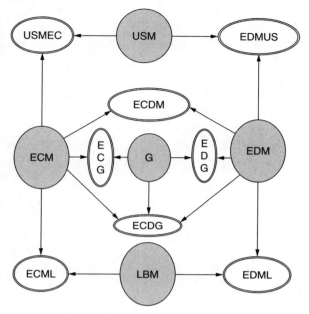

Figure 1.9 Hybrid machining processes.

References

El-Kady, E. Y., Nassef, G. A., and El-Hofy, H. (1998). "Tool Wear Characteristics During Hot Machining," *Scientific Bulletin*, Ain Shams University, 33 (4): 493–511.

Kaczmarek, J. (1976). *Principles of Machining by Cutting, Abrasion, and Erosion.* Stevenage, U.K.: Peter Peregrines, Ltd.

Kozak, J., and Rajurkar, K. P. (2000). "Hybrid Machining Process Evaluation and Development," Keynote Paper, *Second International Conference on Machining and Measurements of Sculptured Surfaces*, Krakow, pp. 501–536.

McGeough, J. A. (1988). Advanced *Methods of Machining.* London, New York: Chapman and Hall.

McGeough, J. A. (2002). *Micromachining of Engineering Materials.* New York: Marcel Dekker, Inc.

Tanigushi, N. (1983). "Current Status in and Future Trends of Ultra Precision Machining and Ultra Fine Materials Processing," *Annals of CIRP*, 32 (2): 573–582.

Todd, J. A., and Copley, S. M. (1997). "Development of a Prototype Laser Processing System for Shaping Advanced Ceramic Material," ASME, *Journal of Manufacturing Science and Engineering*, 119: 55–67.

www.unl.edu.nmrc/outline.htm

Chapter 2

Mechanical Processes

2.1 Ultrasonic Machining

2.1.1 Introduction

Ultrasonic machining (USM) is the removal of hard and brittle materials using an axially oscillating tool at ultrasonic frequencies [18–20 kilohertz (kHz)]. During that oscillation, the abrasive slurry of B_4C or SiC is continuously fed into the machining zone between a soft tool (brass or steel) and the workpiece. The abrasive particles are, therefore, hammered into the workpiece surface and cause chipping of fine particles from it. The oscillating tool, at amplitudes ranging from 10 to 40 µm, imposes a static pressure on the abrasive grains and feeds down as the material is removed to form the required tool shape (Fig. 2.1). Balamuth first discovered USM in 1945 during ultrasonic grinding of abrasive powders. The industrial applications began in the 1950s when the new machine tools appeared. USM is characterized by the absence of any deleterious effect on the metallic structure of the workpiece material.

2.1.2 The machining system

The machining system, shown in Figs. 2.2 and 2.3, is composed mainly from the magnetostrictor, concentrator, tool, and slurry feeding arrangement. The magnetostrictor is energized at the ultrasonic frequency and produces small-amplitude vibrations.

Such a small vibration is amplified using the constrictor (mechanical amplifier) that holds the tool. The abrasive slurry is pumped between the oscillating tool and the brittle workpiece. A static pressure is applied in the tool-workpiece interface that maintains the abrasive slurry.

16 Chapter Two

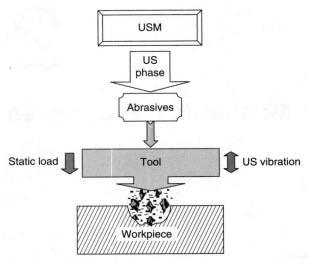

Figure 2.1 USM components.

2.1.2.1 The magnetostrictor. The magnetostrictor used in USM, shown in Fig. 2.4, has a high-frequency winding wound on a magnetostrictor core and a special polarizing winding around an armature. The magnetostriction effect was first discovered by Joule at Manchester in 1874. Accordingly, a magnetic field undergoing ultrasonic frequencies causes

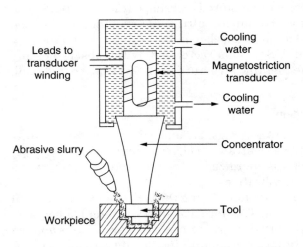

Figure 2.2 Main elements of an ultrasonic machining system.

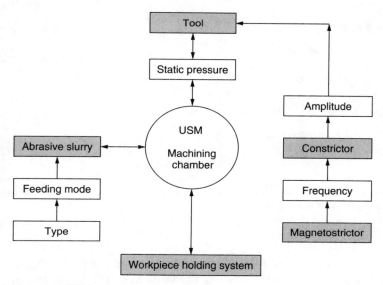

Figure 2.3 USM system components.

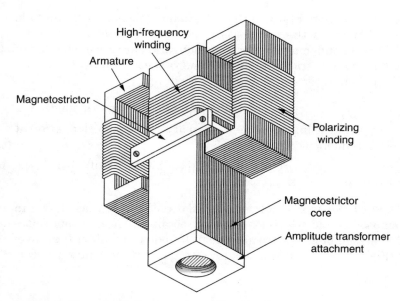

Figure 2.4 Magnetostriction transducer (Kaczmarek, 1976).

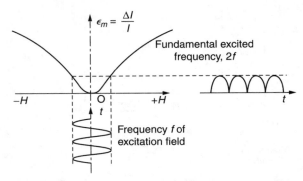

Figure 2.5 Magnetostrictor excited by a variable magnetic field without magnetizing (Kaczmarek, 1976).

corresponding changes in a ferromagnetic object placed within its region of influence. This effect is used to oscillate the USM tool, which is mounted at the end of a magnetostrictor, at ultrasonic frequencies (18 to 20 kHz). The method of operation of a magnetostrictor can be explained as follows.

The coefficient of magnetostriction elongation ϵ_m is

$$\epsilon_m = \frac{\Delta l}{l}$$

where Δl is the incremental length of the magnetostrictor core and l is the original length of the magnetostrictor core, both in millimeters. Materials having high magnetostrictive elongation are recommended to be used for a magnetostrictor. Figure 2.5 shows the relationship between the magnetic field intensity H and ϵ_m. Accordingly,

- The elongation is independent of the sign of the magnetic field.
- The variation of the magnetic field intensity changes in elongation at double the frequency ($2f$).
- Changes in elongation are not sinusoidal (full wave rectified) as is the case for the field intensity.

If the transducer is magnetized with a direct current, as shown in Fig. 2.6, sinusoidal changes in elongation are obtained. The maximum elongation A_{max} in the magnetostrictor of length l equal to half of the wavelength λ (Fig. 2.7) will occur at a distance of $\lambda/4$ from the center. Hence,

$$\lambda = \frac{C_s}{f}$$

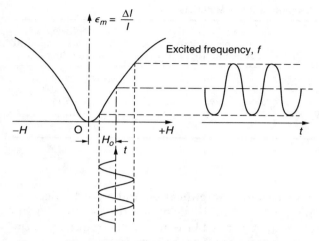

Figure 2.6 Magnetostriction due to a variable magnetic field after polarization (Kaczmarek, 1976).

where C_s is the speed of sound in the magnetostrictor material [meters per second (m/s)] and f is the frequency of the changes in the magnetic field (1/s). Also,

$$\lambda = \frac{1}{f}\sqrt{\frac{E}{P}}$$

where E is Young's modulus [megapascals (Mpa)] and P is the density of the magnetostrictor material [kilograms per cubic meter (kg/m^3)].

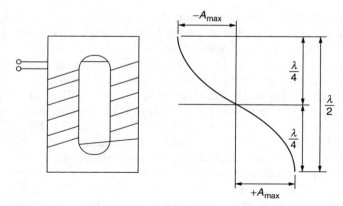

Figure 2.7 Variation in a wave of elongation along the length of the magnetostrictor (Kaczmarek, 1976).

TABLE 2.1 Properties of Magnetostrictive Materials

Material type	Coefficient of magnetostrictive elongation E_m ($\times 10^6$)	Coefficient of magnetomechanical coupling K_m
Alfer (13% Al, 87% Fe)	40	0.28
Hypernik (50% Ni, 50% Fe)	25	0.20
Permalloy (40% Ni, 60% Fe)	25	0.17
Permendur (49% Co, 2% V, 49% Fe)	9	0.20

SOURCE: McGeough (1988).

In order to obtain the maximum amplification and a good efficiency, the magnetostrictor must, therefore, be designed to operate at resonance where its natural frequency must be equal to the frequency of the magnetic field. The resonance frequency f_r becomes

$$f_r = \frac{1}{2l}\sqrt{\frac{E}{P}}$$

Since the magnetostrictor material converts the magnetic energy to a mechanical one, a higher coefficient of magnetomechanical coupling, K_m, is essential.

$$K_m = \sqrt{\frac{E_w}{E_m}}$$

where for magnetostrictive materials, shown in Table 2.1, E_w is the mechanical energy and E_m is the magnetic energy.

2.1.2.2 Mechanical amplifier.

The elongation obtained at the resonance frequency f_r using a magnetostrictor of length $l = 0.5\lambda$ is usually 0.001 to 0.1 μm, which is too small for practical machining applications. The vibration amplitude is increased by fitting an amplifier (acoustic horn) into the output end of the magnetostrictor. Larger amplitudes, typically 40 to 50 μm, are found to be suitable for practical applications. Depending on the final amplitude required, the amplitude amplification can be achieved by one or more acoustic horns (Fig. 2.8). In order to have the maximum amplitude of vibration (resonance) the length of the concentrator is made multiples of one-half the wavelength of sound λ in the concentrator (horn) material. The choice of the shape of the acoustic horn controls the final amplitude. Five acoustic horns (cylindrical, stepped, exponential, hyperbolic cosine, and conical horns) have been reported by Youssef (1976). Exponential and stepped types are frequently used

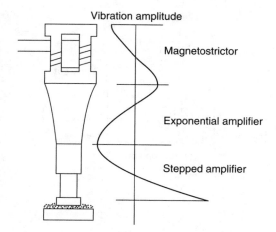

Figure 2.8 Two-step amplification in USM.

because they are easily designed and produced compared to the conical and hyperbolic horns.

Aluminum bronze and marine bronze are cheap with high fatigue strengths of 185 and 150 meganewtons per square meter (MN/m^2), respectively. The main drawbacks of the magnetostrictive transducer are the high losses encountered, the low efficiency (55 percent), the consequent heat up, and the need for cooling. Higher efficiencies (90–95 percent) are possible by using piezoelectric transformers to modern USM machines.

2.1.2.3 Tools. Tool tips must have high wear resistance and fatigue strength. For machining glass and tungsten carbide, copper and chromium silver steel tools are recommended. Silver and chromium nickel steel are used for machining sintered carbides. During USM, tools are fed toward, and held against, the workpiece by means of a static pressure that has to overcome the cutting resistance at the interface of the tool and workpiece. Different tool feed mechanisms are available that utilize pneumatic, periodic switching of a stepping motor or solenoid, compact spring-loaded system, and counterweight techniques as mentioned in *claymore.engineer.gvsu.edu*.

2.1.2.4 Abrasive slurry. Abrasive slurry is usually composed of 50 percent (by volume) fine abrasive grains (100–800 grit number) of boron carbide (B_4C), aluminum oxide (Al_2O_3), or silicon carbide (SiC) in 50 percent water. The abrasive slurry is circulated between the oscillating tool and workpiece.

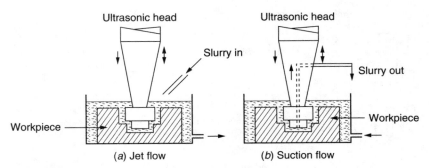

Figure 2.9 Slurry injection methods.

Under the effect of the static feed force and the ultrasonic vibration, the abrasive particles are hammered into the workpiece surface causing mechanical chipping of minute particles. The slurry is pumped through a nozzle close to the tool-workpiece interface at a rate of 25 liters per minute (L/min).

As machining progresses, the slurry becomes less effective as the particles wear and break down. The expected life ranges from 150 to 200 hours (h) of ultrasonic exposure (*Metals Handbook*, 1989). The slurry is continuously fed to the machining zone in order to ensure efficient flushing of debris and keeps the suspension cool during machining. The performance of USM depends on the manner in which the slurry is fed to the cutting zone. Figure 2.9 shows the different slurry feeding arrangements.

2.1.3 Material removal process

Figure 2.10 shows the complete material removal mechanism of USM, which involves three distinct actions:

1. Mechanical abrasion by localized direct hammering of the abrasive grains stuck between the vibrating tool and adjacent work surface.
2. The microchipping by free impacts of particles that fly across the machining gap and strike the workpiece at random locations.
3. The work surface erosion by cavitation in the slurry stream.

The relative contribution of the cavitation effect is reported to be less than 5 percent of the total material removed. The dominant mechanism involved in USM of all materials is direct hammering. Soft and elastic materials like mild steel are often plastically deformed first and are later removed at a lower rate.

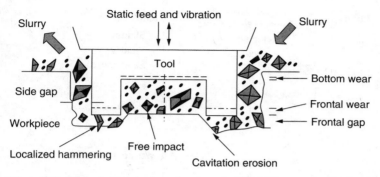

Figure 2.10 Material removal mechanisms in USM (Thoe et al., 1995).

In case of hard and brittle materials such as glass, the machining rate is high and the role played by free impact can also be noticed. When machining porous materials such as graphite, the mechanism of erosion is introduced. The rate of material removal, in USM, depends, first of all, on the frequency of tool vibration, static pressure, the size of the machined area, and the abrasive and workpiece material. The material removal rate and hence the machinability by USM depends on the brittleness criterion which is the ratio of shearing to breaking strength of a material. According to Table 2.2 glass has a higher machinability than that of a metal of similar hardness. Moreover, because of the low brittleness criterion of steel, which is softer, it is used as a tool material. Figure 2.11 summarizes the important parameters that affect the performance of USM, which are mainly related to the tool, workpiece material, abrasives, machining conditions, and the ultrasonic machine (Jain and Jain, 2001).

In USM, the material removal rate (MRR) can generally be described using the following formula (*www2.cerm.wvn.edu/*):

$$\text{MRR} = 5.9 \mathcal{F} \left(\frac{S}{H_0} \right) R^{0.5} Y^{0.5}$$

TABLE 2.2 Relative Machinability Ratings for Some Materials by USM

Work material	Relative removal rate, %
Glass	100
Brass	66
Tungsten	4.8
Titanium	4.0
Steel	3.9
Chrome steel	1.4

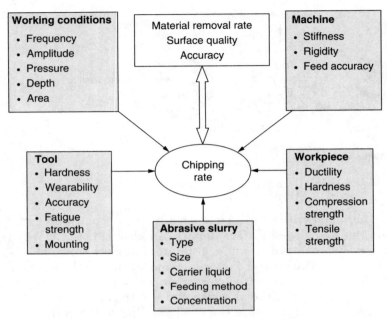

Figure 2.11 Factors affecting USM performance.

where $\mathcal{F}$ = frequency of oscillation
S = static stress on tool, kg/mm^2
H_0 = surface fracture strength, Brinell hardness number (BHN)
R = mean radius of grit, mm
Y = amplitude of vibration, mm

2.1.4 Factors affecting material removal rate

2.1.4.1 Tool oscillation. The amplitude of the tool oscillation has the greatest effect of all the process variables. The material removal rate increases with a rise in the amplitude of the tool vibration. The vibration amplitude determines the velocity of the abrasive particles at the interface between the tool and workpiece. Under such circumstances the kinetic energy rises, at larger amplitudes, which enhances the mechanical chipping action and consequently increases the removal rate. A greater vibration amplitude may lead to the occurrence of splashing, which causes a reduction of the number of active abrasive grains and results in a decrease in the material removal rate.

According to Kaczmarek (1976) with regard to the range of grain sizes used in practice, the amplitude of oscillation varies within the limits of 0.04 to 0.08 mm. The increase of feed force induces greater chipping forces by each grain, which raises the overall removal rate. Regarding the effect of vibration frequency on the removal rate, it has been reported by McGeough (1988) that the increase in vibration frequency reduces the removal rate. This trend may be related to the small chipping time allowed for each grain such that a lower chipping action prevails causing a decrease in the removal rate.

2.1.4.2 Abrasive grains. Both the grain size and the vibration amplitude have a similar effect on the removal rate. According to McGeough (1988), the removal rate rises at greater grain sizes until the size reaches the vibration amplitude, at which stage, the material removal rate decreases. When the grain size is large compared to the vibration amplitude, there is a difficulty of abrasive renewal in the machining gap. Because of its higher hardness, B_4C achieves higher removal rates than silicon carbide (SiC) when machining a soda glass workpiece. The rate of material removal obtained with silicon carbide is about 15 percent lower when machining glass, 33 percent lower for tool steel, and about 35 percent lower for sintered carbide.

Water is commonly used as the abrasive carrying liquid for the abrasive slurry while benzene, glycerol, and oils are alternatives. The increase of slurry viscosity reduces the removal rate. The improved flow of slurry results in an enhanced machining rate. In practice a volumetric concentration of about 30 to 35 percent of abrasives is recommended. A change of concentration occurs during machining as a result of the abrasive dust settling on the machine table. The actual concentration should, therefore, be checked at certain time intervals. The increase of abrasive concentration up to 40 percent enhances the machining rate. More cutting edges become available in the machining zone, which raises the chipping rate and consequently the overall removal rate.

2.1.4.3 Workpiece impact-hardness. The machining rate is affected by the ratio of the tool hardness to the workpiece hardness. In this regard, the higher the ratio, the lower will be the material removal rate. For this reason soft and tough materials are recommended for USM tools.

2.1.4.4 Tool shape. The machining rate is affected by the tool shape and area. An increase in the tool area decreases the machining rate due to the problem of adequately distributing the abrasive slurry over the entire machining zone. It has been reported by McGeough (1988) that, for the same machining area, a narrow rectangular shape yields a higher machining rate than a square cross-sectional one. The rise in the static feed pressure enhances the machining rate up to a limiting condition, beyond which no further increase occurs. The reason behind such a trend is related to the disturbance of the oscillation behavior of the tool at higher forces where lateral vibrations are expected to occur.

According to Kaczmarek (1976), at pressures lower than the optimum, the force pressing the grains into the material is too small and the volume removed by a particular grain diminishes. Beyond the optimum pressure, damping is too strong and the tool ceases to break away from the grains, thus preventing them from changing position, which reduces the removal rate. Measurements also showed a decrease in the material removal rate with an increase in the hole depth. The reason for this is that the deeper the tool reaches, the more difficult and slower is the exchange of abrasives from underneath the tool.

2.1.5 Dimensional accuracy and surface quality

2.1.5.1 Dimensional accuracy. Generally the form accuracy of machined parts suffers from the following disturbing factors, which cause oversize, conicity, and out of roundness.

- Side wear of the tool
- Abrasive wear
- Inaccurate feed of the tool holder
- Form error of the tool
- Unsteady and uneven supply of abrasive slurry around the oscillating tool

Overcut. The process accuracy is measured through the overcut (oversize) produced during drilling of holes. The hole oversize measures the difference between the hole diameter, measured at the top surface, and the tool diameter. The side gap between the tool and the machined hole is necessary to enable the abrasives to flow to the machining zone under the oscillating tool. Hence the grain size of the abrasives represents the main factor, which affects the overcut produced. The overcut is considered to be about two to four times greater than the mean grain size when machining glass and tungsten carbide. It is about three times greater than the mean grain size of B_4C (mesh numbers 280–600). However, the magnitude of the overcut depends on many other process variables including the type of workpiece material and the method of tool feed. In general USM accuracy levels are limited to ±0.05 mm.

Conicity. The overcut is usually greater at the entry side than at the exit one due to the cumulative abrasion effect of the fresh and sharp grain particles. As a result of such an effect, a hole conicity of approximately

0.2° arises when drilling a 20-mm-diameter hole to a depth of 10 mm in graphite. The conicity can be reduced by

- Direct injection of the abrasive slurry into the machining zone
- The use of tools having negatively tapering walls
- The use of high static pressure that produces finer abrasives, which in turn reduces the amount of tool wear and the resulting conicity
- The use of wear-resistant tool materials
- The use of an undersized tool in the first cut and a final tool of the required size, which will cut faster and reduce the conicity

Out of roundness. The out of roundness arises by the lateral vibrations of the tool. Such vibrations may arise due to the out of perpendicularity of the tool face and the tool centerline and when the acoustic parts of the machine are misaligned. Typical roundness errors are about 40 to 140 μm and 20 to 60 μm, respectively, for glass and graphite materials.

2.1.5.2 Surface quality.

The surface finish is closely related to the machining rate in USM. Table 2.3 shows the relationship between grit number and grit size. The larger the grit size, the faster the cutting but the coarser the surface finish. A surface finish of 0.38 to 0.25 μm can be expected using abrasives of grit number 240. However, other factors such as tool surface, amplitude of tool vibration, and material being machined also affect the surface finish. The larger the grit (smaller the grain size), the smoother becomes the produced surface. As mentioned earlier, the larger chipping marks formed on brittle machined materials create rougher surfaces than that obtained in the case of machined hard alloy steel. The amplitude of tool oscillation has a smaller effect on the surface finish. As the amplitude is raised the individual grains are pressed further into the workpiece surface thus causing deeper

TABLE 2.3 Grit Number, Grit Size, and Surface Roughness for USM

Grit number	Grit size, mm	Roughness, μm
180	0.086	0.55
240	0.050	0.51
320	0.040	0.45
400	0.030	0.40
600	0.014	0.28
800	0.009	0.21

craters and hence a rougher surface finish. Other process variables such as static pressure have a little effect on the surface finish.

Smoother surfaces can also be obtained when the viscosity of the liquid carrier of the abrasive slurry is reduced. It is evident that the surface irregularities of the sidewall surfaces of the cavities are considerably larger than those of the bottom. This results from the sidewalls being scratched by grains entering and leaving the machining zone. Cavitation damage to the machined surface occurs when the tool particles penetrate deeper into the workpiece. Under such circumstances it is more difficult to replenish adequately the slurry in these deeper regions and thus a rougher surface is produced.

2.1.6 Applications

USM should be applied for shallow cavities cut in hard and brittle materials having a surface area less than 1000 mm^2.

2.1.6.1 Drilling and coring. A modified version of USM is shown in Fig. 2.12 where a tool bit is rotated against the workpiece in a similar fashion to conventional drilling. The process is, therefore, called rotary ultrasonic machining (RUM). Cruz et al. (1995) used the process for machining nonmetallic materials such as glass, alumina, ceramic, ferrite, quartz, zirconium oxide, ruby, sapphire, beryllium oxide, and some composite materials. RUM ensures high removal rates, lower tool pressures for delicate parts, improved deep hole drilling, less breakout or through holes, and no core seizing during core drilling.

The process allows the uninterrupted drilling of small-diameter holes, while conventional drilling necessitates a tool retraction, which increases the machining time. The penetration rate depends on the size and depth

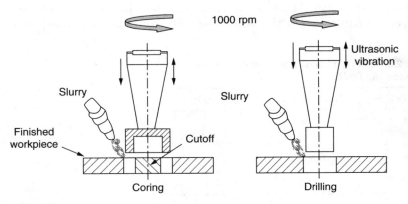

Figure 2.12 Rotary USM.

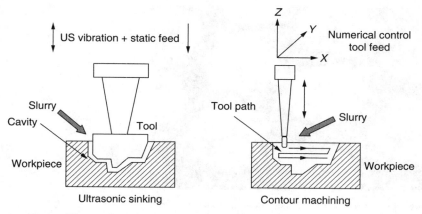

Figure 2.13 Ultrasonic sinking and contour machining.

of the cavity. Small holes require more time as the rate of machining decreases with the depth of penetration due to the difficulty in maintaining a continuous supply of new slurry at the tool face. Generally a depth-to-diameter ratio of 2.5 is achievable by RUM.

2.1.6.2 Ultrasonic sinking and contour machining. During USM sinking, the material removal is difficult when the machined depth exceeds 5 to 7 mm or when the active section of the tool becomes important. Under such conditions the removal of the abrasive grits at the interface becomes difficult and hence the material removal process is impossible. Moreover the manufacture of such a tool is generally complex and costly. Contouring USM (Fig. 2.13) employs simple tools that are moved in accordance to the contour required (Benkirane et al., 1995). Figure 2.14 shows a three-dimensional shape machined by USM sinking where the shaped tool is used to produce a negative replica in the workpiece.

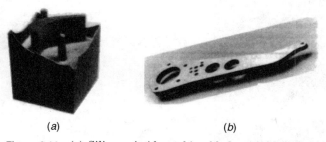

Figure 2.14 (a) Silicon nitride turbine blades (sinking), and (b) CFC acceleration lever and holes (contour USM) (Benkirane et al., 1995).

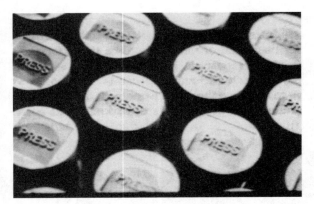

Figure 2.15 Graphite EDM electrodes machined by USM (Gilmore, 1995).

The same figure also shows holes and contours machined using a USM contour machining.

2.1.6.3 Production of EDM electrodes. Gilmore (1995) used USM to produce graphite EDM electrodes as shown in Fig. 2.15. Typical ultrasonic machining speeds, in graphite, range from 0.4 to 1.4 centimeters per minute (cm/min). The surface roughness ranges from 0.2 to 1.5 μm and accuracies of ±10 μm are typical. Small machining forces permit the manufacture of fragile graphite EDM electrodes.

Before

After

Figure 2.16 Ultrasonic polishing of CNC machined parts (Gilmore, 1995).

2.1.6.4 Ultrasonic polishing. Ultrasonic polishing occurs by vibrating a brittle tool material such as graphite or glass into the workpiece at an ultrasonic frequency and a relatively low vibration amplitude. The fine abrasive particles, in the slurry, abrade the high spots of the workpiece surface, typically removing 0.012 mm of material or less. Using such a technique Gilmore (1995) reported the surface finish to be as low as 0.3 µm. Figure 2.16 shows the ultrasonic polishing that lasted 1.5 to 2 min to remove the machining marks left by a computer numerical control (CNC) engraving operation.

2.1.6.5 Micro-ultrasonic machining. Micro-ultrasonic machining (MUSM) is a method that utilizes workpiece vibration. According to Egashira and Masuzana (1999) vibrating the workpiece allows for freer tool system design because it does not include the set of transducer, horn, and cone. In addition, the complete system is much more simple and compact than conventional USM (Fig. 2.17). Using such a method microholes of 5-µm diameter on quartz, glass, and silicon have been produced using tungsten carbide (WC) alloy microtools.

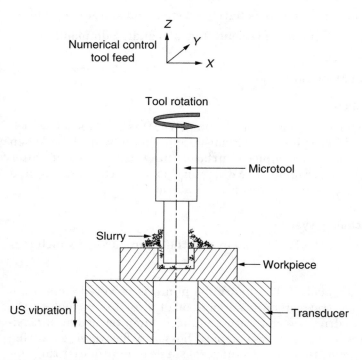

Figure 2.17 Micro-ultrasonic machining.

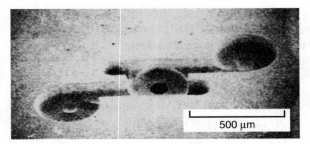

Figure 2.18 Micro-ultrasonic machined cavity (Masuzawa and Tonshof, 1997).

However the high wear resistance of sintered diamond (SD) tools made it possible to machine multiple holes using a single tool. Similarly MUSM is used for machining three-dimensional shapes as shown in Fig. 2.18.

2.1.6.6 Other applications

- Cutting off parts made from semiconductors at high removal rates compared to conventional machining methods
- Engraving on glass as well as hardened steel and sintered carbide
- Parting and machining of precious stones including diamond

2.2 Water Jet Machining

2.2.1 Introduction

The key element in water jet machining (WJM) is a water jet, which travels at velocities as high as 900 m/s (approximately Mach 3). When the stream strikes a workpiece surface, the erosive force of water removes the material rapidly. The water, in this case, acts like a saw and cuts a narrow groove in the workpiece material.

2.2.2 The machining system

Figure 2.19 shows the WJM system and the main parts of which it is composed.

2.2.2.1 Hydraulic pump. The hydraulic pump is powered from a 30-kilowatt (kW) electric motor and supplies oil at pressures as high as 117 bars in order to drive a reciprocating plunger pump termed an *intensifier*. The hydraulic pump offers complete flexibility for water jet cutting and cleaning applications. It also supports single or multiple cutting stations for increased machining productivity.

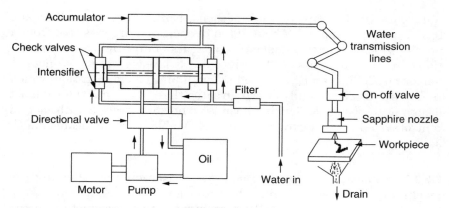

Figure 2.19 Schematic illustration of WJM system.

2.2.2.2 Intensifier. The intensifier accepts the water at low pressure (typically 4 bar) and expels it, through an accumulator, at higher pressures of 3800 bar. The intensifier converts the energy from the low-pressure hydraulic fluid into ultrahigh-pressure water. The hydraulic system provides fluid power to a reciprocating piston in the intensifier center section. A limit switch, located at each end of the piston travel, signals the electronic controls to shift the directional control valve and reverses the piston direction. The intensifier assembly, with a plunger on each side of the piston, generates pressure in both directions. As one side of the intensifier is in the inlet stroke, the opposite side is generating ultrahigh-pressure output. During the plunger inlet stroke, filtered water enters the high-pressure cylinder through the check value assembly. After the plunger reverses direction, the water is compressed and exits at ultrahigh pressure.

2.2.2.3 Accumulator. The accumulator maintains the continuous flow of the high-pressure water and eliminates pressure fluctuations. It relies on the compressibility of water (12 percent at 3800 bar) in order to maintain a uniform discharge pressure and water jet velocity, when the intensifier piston changes its direction.

2.2.2.4 High-pressure tubing. High-pressure tubing transports pressurized water to the cutting head. Typical tube diameters are 6 to 14 mm. The equipment allows for flexible movement of the cutting head. The cutting action is controlled either manually or through a remote-control valve specially designed for this purpose.

2.2.2.5 Jet cutting nozzle. The nozzle provides a coherent water jet stream for optimum cutting of low-density, soft material that is considered unmachinable by conventional methods. Nozzles are normally made from synthetic sapphire. About 200 h of operation are expected from a nozzle, which becomes damaged by particles of dirt and the accumulation of mineral deposits on the orifice due to erosive water hardness. A longer nozzle life can be obtained through multistage filtration, which removes undesired solids of size greater than 0.45 µm. The compact design of the water jet cutting head promotes integration with motion control systems ranging from two-axis (XY) tables to sophisticated multi-axis robotic installations.

2.2.2.6 Catcher. The catcher acts as a reservoir for collecting the machining debris entrained in the water jet. Moreover, it reduces the noise levels [105 decibels (dB)] associated with the reduction in the velocity of the water jet from Mach 3 to subsonic levels.

2.2.3 Process parameters

Jet nozzle. The standoff distance, shown in Fig. 2.20, is the gap between the jet nozzle (0.1–0.3 mm diameter) and the workpiece (2.5–6 mm). However for materials used in printed circuit boards, it may be increased to 13 to 19 mm. For a nozzle of 0.12-mm diameter and cutting rate of 1.1 millimeters per second (mm/s), McGeough (1988) reported the decrease of the depth of cut at a larger standoff distance. When cutting fiber-reinforced plastics, reports showed that the increase in machining rate and use of the small nozzle diameter increased the width of the damaged layer.

Jet fluid. Typical pressures reported by McGeough (1988) are 150 to 1000 MPa, which provide 8 to 80 kW of power. For a given nozzle diameter, the

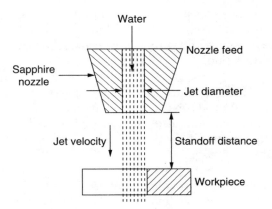

Figure 2.20 WJM terminology.

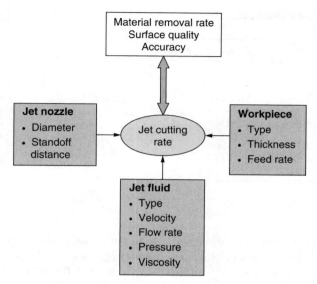

Figure 2.21 Factors affecting WJM performance.

increase in pressure allows more power to be used in the machining process, which in turn increases the depth of the cut. Jet velocities range between 540 to 1400 m/s. The quality of cutting improves at higher pressures by widening the diameter of the jet and by lowering the traverse speed. Under such conditions, materials of greater thicknesses and densities can be cut. Moreover, the larger the pump pressure, the greater will be the depth of the cut. The fluid used must possess low viscosity to minimize the energy losses and be noncorrosive, nontoxic, common, and inexpensive. Water is commonly used for cutting alloy steels. Alcohol is used for cutting meat, while cooking oils are recommended for cutting frozen foods. Figure 2.21 summarizes different parameters affecting the performance of WJM.

Target material. Brittle materials will fracture, while ductile ones will cut well. Material thicknesses range from 0.8 to 25 mm or more. Table 2.4 shows the cutting rates for different material thicknesses.

2.2.4 Applications

WJM is used on metals, paper, cloth, leather, rubber, plastics, food, and ceramics. It is a versatile and cost-effective cutting process that can be used as an alternative to traditional machining methods. It completely eliminates heat-affected zones, toxic fumes, recast layers, work hardening, and thermal stresses. It is the most flexible and effective cleaning

TABLE 2.4 Typical Water Jet Cutting Rates for Various Materials

Material	Thickness, mm	Feed rate, m/min
Leather	2.2	20
Vinyl chloride	3.0	0.5
Polyester	2.0	150
Kevlar	3.0	3
Graphite	2.3	5
Gypsum board	10	6
Corrugated board	7	200
Pulp sheet	2	120
Plywood	6	1

SOURCE: Tlusty (1999).

solution available for a variety of industrial needs. In general the cut surface has a sandblast appearance. Moreover, harder materials exhibit a better edge finish. Typical surface finishes ranges from 1.6 µm root mean square (RMS) to very coarse depending on the application. Tolerances are in the range of ±25 µm on thin material. Both the produced surface roughness and tolerance depend on the machining speed.

2.2.4.1 Cutting. WJM is limited to fiberglass and corrugated wood. Figure 2.22 shows a typical example of water jet cutting of marble, while Fig. 2.23 shows a typical application in the food industry.

2.2.4.2 Drilling. The process drills precision-angled and -shaped holes in a variety of materials for which other processes such as EDM or EBM are too expensive or too slow.

2.2.4.3 Machining of fiber-reinforced plastics. In this case the thermal material damage is negligible. The tool, being effectively pointed, accurately cuts contours. The main drawback is the deflection of the water

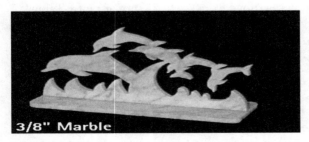

Figure 2.22 Water jet cutting example (*www.jetcutinc.com/*).

Figure 2.23 Water jet cutting of tomatoes (*www.jetedge.com/*).

jet by the fiber embedded in the matrix, which protrudes after machining. The feed rate attainable depends on the surface quality required. Table 2.5 shows the limiting feed rates for water jet cutting of fiber-reinforced plastics.

2.2.4.4 Cutting of rocks. Water jet cutting of a 51-mm-deep slot in granite using two oscillating jets at 275 MPa during 14 passes at a 25.4-mm/s feed rate has been reported by McGeough (1988). Moreover an oscillating nozzle system operating at the same feed rate and pressure of 172 MPa, with the standoff distance adjusted every pass was used to cut a 178-mm-deep slot in sandstone.

2.2.4.5 Deburring. The method uses large pressures to remove large burrs (3 mm height) in 12-mm-diameter drilled holes in a hollow molybdenum-chromium steel shaft at 15 s using 700-bar pressure and a flow rate of 27 L/min. In this method burrs are broken off by the

TABLE 2.5 Limiting Feed Rates for WJC of Fiber-Reinforced Plastics with 3500-bar Pressure, 0.1-mm Nozzle, and 2-mm Distance

Material	Thickness, mm	Feed rate, m/min
Glass fiber-reinforced polymers (GFRP) (laminate)	2.2	1.8–6.0
	3.0	1.4–5.0
	5.0	0.7–6.0
Aramid fiber-reinforced polymers (AFRP) (weave)	1.0	10.0
	2.0	2.4–4.0

SOURCE: McGeough (1988).

impact of water. A higher pressure (4000 bar) and a lower flow rate (2.5 L/min) are used to remove burrs from nonmetallic materials.

2.2.4.6 Cutting of printed circuit boards. Using a small-diameter water jet mounted near to the part edge, a printed circuit board (PCB) can be cut at a speed that exceeds 8 m/min, to the accuracy of ±0.13 mm. Boards of various shapes for use in portable radios and cassette players can be cut using computer numerical control (CNC) technology.

2.2.4.7 Surface treatment. The process finds many applications including:

- Removing deposits and residues without toxic chemicals, which eliminates costly cleanup and disposal problems
- Surface cleaning of pipes and castings, decorative finishing, nuclear decontamination, food utensil cleaning, degreasing, polishing, preparation for precise inspection, and surface texturing
- Economical surface preparation and coating removal
- Removing corrosion, spray residue, soluble salts, chemicals, and surface damage prior to recoating or painting

2.2.4.8 Wire stripping. The process can remove the wire insulating material without damaging the metal or removing the tinning on the copper wire. The processing time can be decreased to about 20 percent of the manual stripping method (*Metals Handbook*, 1989).

2.2.5 Advantages and disadvantages of WJM

Advantages

- It has multidirectional cutting capacity.
- No heat is produced.
- Cuts can be started at any location without the need for predrilled holes.
- Wetting of the workpiece material is minimal.
- There is no deflection to the rest of the workpiece.
- The burr produced is minimal.
- The tool does not wear and, therefore, does not need sharpening.
- The process is environmentally safe.
- Hazardous airborne dust contamination and waste disposal problems that are common when using other cleaning methods are eliminated.

- There is multiple head processing.
- Simple fixturing eliminates costly and complicated tooling, which reduces turnaround time and lowers the cost.
- Grinding and polishing are eliminated, reducing secondary operation costs.
- The narrow kerf allows tight nesting when multiple parts are cut from a single blank.
- It is ideal for roughing out material for near net shape.
- It is ideal for laser reflective materials such as copper and aluminum.
- It allows for more accurate cutting of soft material.
- It cuts through very thick material such as 383 mm in titanium and 307 mm in Inconel.

Disadvantages

- Hourly rates are relatively high.
- It is not suitable for mass production because of high maintenance requirements.

2.3 Abrasive Jet Machining

2.3.1 Introduction

In abrasive jet machining (AJM) a focused stream of abrasive grains of Al_2O_3 or SiC carried by high-pressure gas or air at a high velocity is made to impinge on the work surface through a nozzle of 0.3- to 0.5-mm diameter. The process differs from sandblasting (SB) in that AJM has smaller-diameter abrasives and a more finely controlled delivery system. The workpiece material is removed by the mechanical abrasion (MA) action of the high-velocity abrasive particles. AJM machining is best suited for machining holes in superhard materials. It is typically used to cut, clean, peen, deburr, deflash, and etch glass, ceramics, or hard metals.

2.3.2 Machining system

In the machining system shown in Fig. 2.24, a gas (nitrogen, CO_2, or air) is supplied under a pressure of 2 to 8 kg/cm^2. Oxygen should never be used because it causes a violent chemical reaction with workpiece chips or abrasives. After filtration and regulation, the gas is passed through a mixing chamber that contains abrasive particles and vibrates at 50 Hz. From the mixing chamber, the gas, along with the entrained abrasive particles (10–40 µm), passes through a 0.45-mm-diameter tungsten carbide

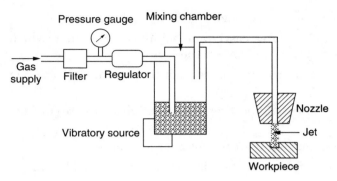

Figure 2.24 AJM system.

nozzle at a speed of 150 to 300 m/s. Aluminum oxide (Al_2O_3) and silicon carbide powders are used for heavy cleaning, cutting, and deburring. Magnesium carbonate is recommended for use in light cleaning and etching, while sodium bicarbonate is used for fine cleaning and the cutting of soft materials. Commercial-grade powders are not suitable because their sizes are not well classified. They may contain silica dust, which can be a health hazard. It is not practical to reuse the abrasive powder because contaminations and worn grit will cause a decline of the machining rate. The abrasive powder feed rate is controlled by the amplitude of vibrations in the mixing chamber. The nozzle standoff distance is 0.81 mm. The relative motion between the workpiece and the nozzle is manually or automatically controlled using cam drives, pantographs, tracer mechanisms, or using computer control according to the cut geometry required. Masks of copper, glass, or rubber may be used to concentrate the jet stream of abrasive particles to a confined location on the workpiece. Intricate and precise shapes can be produced by using masks with corresponding contours. Dust removal equipment is incorporated to protect the environment.

2.3.3 Material removal rate

As shown in Fig. 2.25, the abrasive particles from the nozzle follow parallel paths for a short distance and then the abrasive jet flares outward like a narrow cone. When the sharp-edged abrasive particles of Al_2O_3 or SiC hit a brittle and fragile material at high speed, tiny brittle fractures are created from which small particles dislodge. The lodged out particles are carried away by the air or gas. The material removal rate VRR, is given by

$$\text{VRR} = KN d_a^3 v^{3/2} \left(\frac{\rho_a}{12 H_w} \right)^{3/4}$$

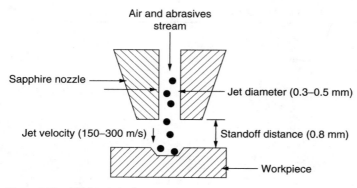

Figure 2.25 AJM terminology.

where K = constant
  N = number of abrasive particles impacting/unit area
  d_a = mean diameter of abrasive particles, μm
  ρ_a = density of abrasive particles, kg/mm^3
  H_w = hardness number of the work material
  v = speed of abrasive particles, m/s

The material removal rate, cut accuracy, surface roughness, and nozzle wear are influenced by the size and distance of the nozzle; composition, strength, size, and shape of abrasives; flow rate; and composition, pressure, and velocity of the carrier gas. The material removal rate is mainly dependent on the flow rate and size of abrasives. Larger grain sizes produce greater removal rates. At a particular pressure, the volumetric removal rate increases with the abrasive flow rate up to an optimum value and then decreases with any further increase in flow rate. This is due to the fact that the mass flow rate of the gas decreases with an increase in the abrasive flow rate and hence the mixing ratio increases causing a decrease in the removal rate because of the decreasing energy available for material removal.

The typical material removal rate is 16.4 mm^3/min when cutting glass. Cutting rates for metals vary from 1.6 to 4.1 mm^3/min. For harder ceramics, cutting rates are about 50 percent higher than those for glass. The minimum width of cut can be 0.13 mm. Tolerances are typically ±0.13 mm with ±0.05 mm possible using good fixation and motion control. The produced surface has a random or matte texture. Surface roughnesses of 0.2 to 1.5 μm using 10 and 50 μm particles, respectively, can be attained. Taper is present in deep cuts. High nozzle pressures result in a greater removal rate, but the nozzle life is decreased. Table 2.6 summarizes the overall process characteristics.

TABLE 2.6 AJM Process Characteristics

Abrasives	
Type	Al_2O_3 or SiC (used once)
Size	Around 25 μm
Flow rate	3–20 g/min
Medium	
Type	Air or CO_2
Velocity	150–300 m/s
Pressure	2–8 kg/cm^2
Flow rate	28 L/min
Nozzle	
Material	Tungsten carbide or sapphire
Shape	Circular, 0.3–0.5 mm diameter
	Rectangular (0.08 × 0.51 mm to 6.61 × 0.51 mm)
Tip distance	0.25–15 mm
Life	WC (12–30 h), sapphire (300 h)
Operating angle	Vertical to 60° off vertical
Area	0.05–0.2 mm^2
Tolerance	±0.05 mm
Surface roughness	0.15–0.2 μm (10-μm particles)
	0.4–0.8 μm (25-μm particles)
	1.0–1.5 μm (20-μm particles)

2.3.4 Applications

1. Drilling holes, cutting slots, cleaning hard surfaces, deburring, polishing, and radiusing
2. Deburring of cross holes, slots, and threads in small precision parts that require a burr-free finish, such as hydraulic valves, aircraft fuel systems, and medical appliances
3. Machining intricate shapes or holes in sensitive, brittle, thin, or difficult-to-machine materials
4. Insulation stripping and wire cleaning without affecting the conductor
5. Micro-deburring of hypodermic needles
6. Frosting glass and trimming of circuit boards, hybrid circuit resistors, capacitors, silicon, and gallium
7. Removal of films and delicate cleaning of irregular surfaces because the abrasive stream is able to follow contours

2.3.5 Advantages and limitations of AJM

Advantages

- Because AJM is a cool machining process, it is best suited for machining brittle and heat-sensitive materials like glass, quartz, sapphire, and ceramics.

- The process is used for machining superalloys and refractory materials.
- It is not reactive with any workpiece material.
- No tool changes are required.
- Intricate parts of sharp corners can be machined.
- The machined materials do not experience hardening.
- No initial hole is required for starting the operation as required by wire EDM.
- Material utilization is high.
- It can machine thin materials.

Limitations

- The removal rate is slow.
- Stray cutting can't be avoided (low accuracy of ±0.1 mm).
- The tapering effect may occur especially when drilling in metals.
- The abrasive may get impeded in the work surface.
- Suitable dust-collecting systems should be provided.
- Soft materials can't be machined by the process.
- Silica dust may be a health hazard.
- Ordinary shop air should be filtered to remove moisture and oil.

2.4 Abrasive Water Jet Machining

2.4.1 Introduction

WJM is suitable for cutting plastics, foods, rubber insulation, automotive carpeting and headliners, and most textiles. Harder materials such as glass, ceramics, concrete, and tough composites can be cut by adding abrasives to the water jet during abrasive water jet machining (AWJM), which was first developed in 1974 to clean metal prior to surface treatment of the metal. The addition of abrasives to the water jet enhanced the material removal rate and produced cutting speeds between 51 and 460 mm/min. Generally, AWJM cuts 10 times faster than the conventional machining methods of composite materials. Zheng et al. (2002) claimed that the abrasive water jet is hundreds, if not thousands, of times more powerful than the pure water jet.

AWJM uses a low pressure of 4.2 bar to accelerate a large volume of a water (70 percent) and abrasive (30 percent) mixture up to a velocity of 30 m/s. Silicon carbides, corundum, and glass beads of grain size

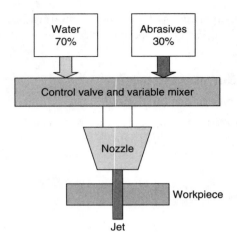

Figure 2.26 AWJM elements.

10 to 150 µm are often used as abrasive materials (Fig. 2.26). Using such a method, burrs of 0.35 mm height and 0.02 mm width left in steel component after grinding are removed by the erosive effect of the abrasives while water acts as an abrasive carrier that dampens its impact effect on the surface. The introduction of compressed air to the water jet enhances the deburring action.

2.4.2 The machining system

In AWJM, the water jet stream accelerates abrasive particles, not the water, to cause the material removal. After the pure water jet is created, abrasives are added using either the injection or suspension methods shown in Fig. 2.27. The important parameters of the abrasives are the material structure and hardness, the mechanical behavior, grain shape, grain size, and distribution.

The basic machining system of AWJM incorporates the following elements.

- Water delivery
- Abrasive hopper and feeder
- Intensifier
- Filters
- Mixing chamber
- Cutting nozzles
- Catcher

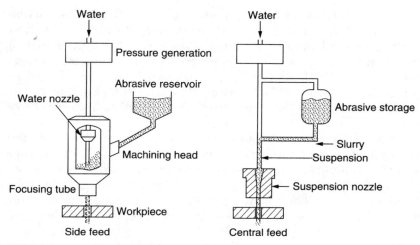

Figure 2.27 Injection and suspension jets.

2.4.3 Process capabilities

Typical process variables include pressure, nozzle diameter, standoff distance, abrasive type, grit number, and workpiece feed rate. An abrasive water jet cuts through 356.6-mm-thick slabs of concrete or 76.6-mm-thick tool steel plates at 38 mm/min in a single pass. The produced surface roughness ranges between 3.8 and 6.4 µm, while tolerances of ±0.13 mm are obtainable. Repeatability of ±0.04 mm, squareness of 0.043 mm/m, and straightness of 0.05 mm per axis are expected. Foundry sands are frequently used for cutting of gates and risers. However, garnet, which is the most common abrasive material, is 30 percent more effective than sand.

During machining of glass a cutting rate of 16.4 mm^3/min is achieved, which is 4 to 6 times that for metals. Surface roughness depends on the workpiece material, grit size, and type of abrasives. A material with a high removal rate produces large surface roughness. For this reason, fine grains are used for machining soft metals to obtain the same roughness as hard ones. The decrease of surface roughness, at a smaller grain size, is related to the reduced depth of cut and the undeformed chip cross section. In addition the larger the number of grains per unit slurry volume, the more that fall on a unit surface area.

A carrier liquid consisting of water with anticorrosive additives has a much greater density than air. This contributes to higher acceleration of the grains with a consequent higher grain speed and increased metal removal rate. Moreover, the carrier liquid spreads over the surface filling its cavities and forming a film that impedes the striking action of the grains. Bulges and the tops of surface irregularities are the first to be affected, and the surface quality improves. Kaczmarek (1976) showed that the use of water air jet permits one to obtain, on average, a roughness number higher by one, as compared with the effect of an air jet. In high-speed WJM of Inconel, Hashish (1992) concluded that the roughness increases at higher feed rates as well as at lower slurry flow rates.

Advanced water jet and AWJ machines are now available where the computer loads a computer-aided design (CAD) drawing from another system. The computer determines the starting and end points and the sequence of operations. The operator then enters the material type and tool offset data. The computer determines the feed rate and performs cutting. Other machining systems operate with a modem and CAD/computer-aided manufacturing (CAM) capabilities that permits transfer from CATIA, AUTOCAD, IGES, and DXF formats. The computer runs a program that determines, in seconds, how to minimize the waste when cutting from blocks or plates (*www.jetedge.com/*).

2.5 Ice Jet Machining

2.5.1 Introduction

The main drawback of WJM is the low efficiency of energy transfer between the jet and the workpiece. This produces low cutting rates, which limits the use of the water jet for machining of comparatively soft materials. For any engineering material, AWJM can be employed. However, the energy efficiency of AWJM is still low. Mixing of water and abrasives limits the minimum jet diameter that can be used.

2.5.2 Process description

In ice jet machining (IJM), the abrasives are replaced by ice particles that form the ice jet. Since the hardness of the ice particles are less than that of the abrasives, lower material removal rates are expected, compared to AWJM. However, the cost reduction and the good environmental impacts make IJM even better. IJM is used in the food, electronic, medical, and space industries where contamination is impossible.

Ice particles are produced using either stream freezing (<500 μm) or ice particles supply (>500 μm) into the stream. In the latter case ice cubes, supplied from an icemaker, are fed to a grinder. Solid CO_2 is added to prevent the crushed ice from melting as shown in Fig. 2.28. The crushed ice is then fed through the machining nozzle. Prior to the nozzle

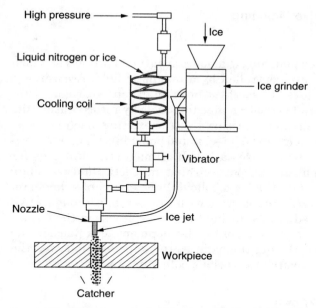

Figure 2.28 IJM schematic, modified from Geskin et al. (1995).

the water is also cooled by passing through a coil that is submerged in liquid nitrogen. Geskin et al. (1995) reported a substantial improvement in the machining characteristics due to the entrapment of ice in the cutting nozzle (see Table 2.7).

TABLE 2.7 Comparison between Water Jet (WJ) and Ice Jet (IJ) Drilling with 320-MPa Pressure, 0.175-mm Nozzle

Material	Thickness, mm	Depth, mm		Diameter, mm		Time, min	Removal rate, mm^3/min		Particle size
		WJ	IJ	WJ	IJ		WJ	IJ	
Aluminum	20	4.3	10.5	1.4	2.8	2.0	3.31	32.3	Large
Steel	6.4	2.5	6.4	1.1	1.1	5.6	0.43	1.09	Large
Steel	2.9	2.1	2.9	1.1	1.1	2.1	0.95	1.31	Small
Ti alloy	12.9	3.1	4.3	1.1	1.2	4.2	0.70	1.16	Small
Graphite	7.4	5.3	7.4	1.2	1.2	5.0	6.90	9.63	Large
Stainless steel	3.2	2.5	2.9	1.1	1.1	3.0	0.79	0.92	Small
Stainless steel	2	0.0	1.7	1.1	1.1	8.0	0.00	0.20	Large

SOURCE: Data from Geskin et al. (1995).

2.6 Magnetic Abrasive Finishing

2.6.1 Introduction

Magnetic field–assisted polishing is a nonconventional process in which the machining forces are controlled by a magnetic field. Accordingly, finish polishing is achieved without the need for expensive, rigid, ultraprecision, vibration- and error-free machine tools by incorporating the magnetic polishing elements necessary into the existing machine tools. There are two types of magnetic field–assisted polishing: magnetic abrasive finishing (MAF), which uses a brush of magnetic abrasives for finish machining, and magnetic float polishing (magnetic fluid grinding), which uses magnetic fluid that is a colloidal dispersion of subdomain magnetic particles in a liquid carrier with abrasives. Although MAF originated in the United States during the forties, it was in the former U.S.S.R. and Bulgaria that much of the development took place in the late fifties and sixties. During the eighties the Japanese followed the work and conducted research for various polishing applications.

2.6.2 The machining system

Figure 2.29 shows a schematic diagram of MAF apparatus. A cylindrical workpiece is clamped into the chuck of the spindle that provides the rotating motion. The workpiece can be a magnetic (steel) or a nonmagnetic (ceramic) material; the magnetic field lines go through the workpiece. Axial vibratory motion is introduced in the magnetic field by the oscillating motion of the magnetic poles relative to the workpiece. A mixture of fine abrasives held in a ferromagnetic material (magnetic abrasive conglomerate, Fig. 2.30) is introduced between the workpiece and the magnetic heads where the finishing process is exerted by the magnetic field. Typically the sizes of the magnetic abrasive conglomerates are 50 to 100 microns and the abrasives are in the 1 to 10 micron

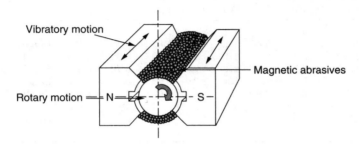

Figure 2.29 MAF schematic.

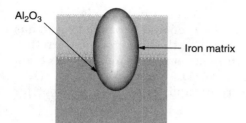

Figure 2.30 Typical magnetic abrasive conglomerates.

range. With nonmagnetic work materials, the magnetic abrasives are linked to each other magnetically between the magnetic N and S poles along the lines of the magnetic forces, forming flexible magnetic abrasive brushes.

In order to achieve uniform circulation of the abrasives, the magnetic abrasives are stirred periodically. Fox et al. (1994) adopted the following MAF conditions that caused both surface and edge finishing:

Roller speed	Up to 1.3 m/s
Magnetic field density	0–0.53 Tesla (T)
Magnetic pressure	0–30 kPa
Abrasive type	80% Fe (40) + 20% SiC (1200)
Vibration frequency	12–25 Hz
Lubricant	Dry or oil

2.6.3 Material removal process

MAF operates with magneto abrasive brushes where the abrasive grains arrange themselves with their carrying iron particles to flexibly comply with the contour of the work surface. The abrasive particles are held firmly against the work surface, while short stroke oscillatory motion is carried out in the axial workpiece direction. MAF brushes contact and act upon the surface protruding elements that form the surface irregularities. While surface defects such as scratches, hard spots, lay lines, and tool marks are removed, form errors like taper, looping, and chatter marks can be corrected with a limited depth of 20 microns. The material removal rate and surface finish depend on the workpiece circumferential speed, magnetic flux density, working clearance, workpiece material, size of magnetic abrasive conglomerates including the type of abrasives used, and its grain size and volume fraction in the conglomerate. Fox et al. (1994) concluded that the average surface finish R_a of a ground rod can be finished to about 10 nm. Increasing the magnetic flux density raises the rate of finishing. High removal rates and the best finish were obtained with an increase in the axial vibration amplitude and frequency. The axial vibration and rotational speed has to be taken into consideration for obtaining the best cross pattern that would give the best finish and high removal rate. Singh and his team (2004) recommended a high voltage level (11.5 V), low working gap (1.25 mm), high rotational speed (180 rpm), and large mesh number for improving the surface quality.

2.6.4 Applications

2.6.4.1 Polishing of balls and rollers. Conventional finishing of ceramic balls, for bearing applications, uses low polishing speeds and diamond abrasives as a polishing medium. The long processing time and the use of expensive diamond abrasives result in high processing costs. Diamond abrasives at high loads can result in deep pits, scratches, and microcracks. Consequently the high processing cost and the lack of the machining system reliability form possible limitations. To minimize the surface damage, gentle polishing conditions are required, namely, low levels of controlled force and abrasives not much harder than the work material.

A recent development in MAF involves the use of a magnetic field to support abrasive slurries in polishing ceramic balls and bearing rollers (Fig. 2.31). A magnetic field, containing abrasive grains and extremely fine ferromagnetic particles in a certain fluid such as water or kerosene, fills the chamber within a guide ring. The ceramic balls are between a drift shaft and a float.

The abrasive grains, ceramic balls, and the float (made from nonmagnetic material) are suspended by the magnetic forces. The balls are preset against the rotating drive shaft and are polished by the mechanical abrasion action. Since the forces applied by the abrasive grains are

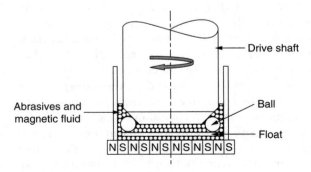

Figure 2.31 Magnetic finishing of balls (Kalpakjian, 1997).

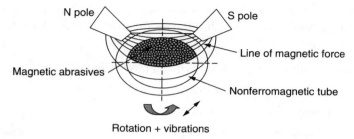

Figure 2.32 Magnetic finishing of nonmagnetic tubes.

extremely small and controllable, the polishing action is very fine. The process is economical, and the surfaces produced have little or no defects.

2.6.4.2 Finishing of inner tube surface. Clean gas and liquid piping systems need to have highly finished inner surfaces that prevent contaminant from accumulating. When the pipe is slender, it is hard to produce smooth inner surfaces in a cost-effective way. Electrolytic finishing has many problems associated with the high cost of controlling the process conditions and disposing of electrolyte without environmental pollution. Figure 2.32 shows the two-dimensional schematic view of the internal finishing of a nonferromagnetic tube using MAF. The magnetic abrasives, inside the tubes, are converged toward the finishing zone by the magnetic field, generating the magnetic force needed for finishing. By rotating the tube at a higher speed, the magnetic abrasives make the inner surface smoother. Figure 2.33 shows the case of ferromagnetic tube finishing where the magnetic fluxes mostly flow into the tube (instead of through the inside of the tube) due to their high magnetic permeability. Under such conditions, the abrasives hardly remain in the finishing zone when the tube is rotated. Geskin et al. (1995) achieved mirror finishing and removed burrs without lowering the accuracy of the shape.

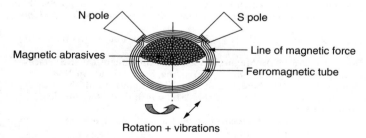

Figure 2.33 Magnetic finishing of magnetic tubes.

2.6.4.3 Other MAF applications. The process can be applied in many other fields, as described by Khayry (2000), Umehara et al. (1997), and Hitomi and Shinmura (1995):

1. Polishing of fine components such as printed circuit boards
2. The removal of oxide layers and protective coatings
3. Chamfering and deburring of gears and cams
4. Automatic polishing of complicated shapes
5. Polishing of flat surfaces

For more details, see the following Internet sites:

www.iijnet.or.jp/MMC/nv.19/fig31.gif

www.jstp,or/jstp_E/Publs/paper/memboo7.html

khai.itl.net.ua/eng/ttc/204/

www.manufacturingcenter.com/tooling/archives/0600/

www.riken.go.jp/lab-www/library/publication/review/

References

Benkirane, Y., Kamoun, H., and Kremer, D. (1995). "Investigation on Ultrasonic Abrasive Material Removal Mechanisms—Analytical and Experimental Study," *Int. Symp. for Electro Machining XI*, Lausanne, Switzerland, pp. 891–900.
catt.bus.okstate.edu/catt2/projects/PhaseIII/finis
claymore.engineer.gvsu.edu/~jackh/eod/manufact/abrasiv/abrasive-2-html#pgfld524546
claymoreengineer.gvsu-edu/~Jackh/eod/manufact/abra.
Cruz, C., Kozak, J., and Rajurkar, K. P. (1995). "Study of Rotary Ultrasonic Machining of Cryogenically Treated Ceramics," *Int. Symp. for Electro Machining XI*, Lausanne, Switzerland, pp. 911–920.
Egashira, K., and Masuzawa, T. (1999). "Micro Ultrasonic Machining by the Application of Workpiece Vibration," *Annals of CIRP*, 48(1):131–134.
El-Hofy, H. (1996). "Surface Generation in Non-conventional Machining," *MDP-6 Conf.*, Cairo, pp. 203–213.
Fox, M., Agrwal, A., Shinmura, T., and Komanduri, R. (1994). "Magnetic Finishing of Rollers," *Annals of CIRP*, 43(1):181–184.
Geskin, E. S., Tismentsky, L., Bokhroi, E., and Li, F. (1995). "Investigation of Ice Jet Machining, *Int. Symp. for Electro Machining XI*, Lausanne, Switzerland, pp. 883–890.
Gilmore, R. (1995). "Ultrasonic Machining and Polishing," *Int. Symp. for Electro Machining XI*, Lausanne, Switzerland, pp. 941–951.
Hashish, M. (1992). "Machining with High Velocity Water Jet," *PEDAC-5 Conf.*, Alexandria, pp. 461–471.
Hitomi, Y., and Shinmura, T. (1995). "Magnetic Abrasive Finishing of Inner Surfaces of Tubes," *Int. Symp. for Electro Machining XI*, Lausanne, Switzerland, pp. 963–976.
Jain, N. K., and Jain, V. K. (2001). "Modeling of Material Removal in Mechanical Type Advanced Machining Processes, a State of the Art," *Journal of Machine Tools and Manufacture*, 41:573–635.
Kaczmarek, J. (1976). *Principles of Machining by Cutting, Abrasion and Erosion.* Stevenage, U.K.: Peter Peregrines, Ltd.

Kalpakjian, S. (1997). *Manufacturing Process for Engineering Materials*. Reading; MA: Addison-Wesley.
khai.itl.net.ua/eng/ttc/204/
Khayry, A. B. (2000). "Aspect of Surface and Edge Finishing by Magneto Abrasive Particles." *Second Int. Conf. on Advanced Manufacturing Technology*, Malaysia, pp. 77–83.
Masuzawa, T., and Tonshof, H. K. (1997). "Three-Dimensional Micro Machining by Machine Tools," *Annals of CIRP*, 46(2):821–828.
McGeough, J. (1988). *Advanced Methods of Machining*. London, New York: Chapman and Hall.
Metals Handbook. (1989). Vol. 16, *Machining*. Materials Park; OH: ASM International.
Singh, D. K., Jain, V. K., and Raghuram, V. (2004). "Parametric Study of Magnetic Abrasive Finishing (MAF) Process," *Int. Symp. for Electro Machining XIV*, On-site Conf. Proc. Edinburgh, U.K. (*Journal of Materials Processing Technology*).
Thoe, T. B., Aspinwal, D. K., and Wise, M. L. (1995). "Towards Ultrasonic Contour Machining," *Int. Symp. for Electro Machining XI*, Lausanne, Switzerland, pp. 953–962.
Umehara, N., Kato, K., and Suziki, K. (1997). "Magnetic Dispersion of Micro Particles Using Magnetic Fluid Application to Texturing Process for Magnetic Rigid Disc," *Annals of CIRP*, 46(1):155–158.
www.jetcutinc.com/
www.jetedge.com/
www.okstate.edu/MAE/maerl/proj2.htm
www.riken.go.jp/lab-www/library/publication/review/
www.manufacturingcenter.com/tooling/archives/0600/
www.iijnet.or.jp/MMC/nv.19/fig31.gif
www.jstp,or/jstp_E/Publs/paper/memboo7.html
www.nedians.8m.com/
www2.cerm.wvn.edu/~imse304/raghav/ragav.htm
Youssef, H. (1976). *Theory of Metal Cutting*, Dar-El-Maaref, Egypt.
Zheng, X., Chen, E., Steele, P., and Grothers, P. (2002). "Shape Machining of Aerospace Composite Components Using Not-Traditional Abrasive Waterjet Cutting Process," *Sixth AMST'02 conference*, Italy, pp. 507–514.

Chapter 3

Chemical Processes

3.1 Chemical Milling

3.1.1 Introduction

Chemical milling (CHM) is the controlled chemical dissolution (CD) of the workpiece material by contact with a strong reagent. Special coatings called *maskants* protect areas from which the metal is not to be removed. The process is used to produce pockets and contours and to remove materials from parts having a high strength-to-weight ratio. CHM consists of the following steps:

1. Preparing and precleaning the workpiece surface. This provides good adhesion of the masking material and assures the absence of contaminants that might interfere with the machining process.
2. Masking using readily strippable mask, which is chemically impregnable and adherent enough to stand chemical abrasion during etching.
3. Scribing of the mask, which is guided by templates to expose the areas that receive CHM. The type of mask selected depends on the size of the workpiece, the number of parts to be made, and the desired resolution of details. Silk-screen masks are preferred for shallow cuts requiring close dimensional tolerances.
4. The workpiece is then etched and rinsed, and the mask is removed before the part is finished.

During CHM (Fig. 3.1), the depth of the etch is controlled by the time of immersion. In order to avoid uneven machining, the chemicals that impinge on the surface being machined should be fresh. The chemicals used are very corrosive and, therefore, must be handled with adequate safety precautions. Both the vapors and the effluents must be suitably

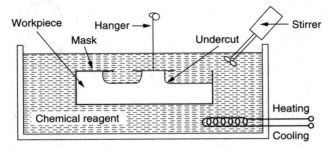

Figure 3.1 CHM setup.

controlled for environmental protection. Agitation of the workpiece and fluid is usual; however, excessive solution flow may result in channeling, grooves, or ridges. Inclination of the workpiece may prevent channeling from gas bubbles. Bellows (1977) and the *Metals Handbook* (1989) reported that dishing of the machined surface occurs due to the uneven heat distribution resulting from the chemical action. Typical reagent temperatures range from 37 to 85°C. Faster etching rates occur at higher temperatures, but must be controlled within ±5°C of the desired temperature in order to attain uniform machining.

When the mask is used, the machining action proceeds both inwardly from the mask opening and laterally beneath the mask thus creating the etch factor shown in Fig. 3.2. The etch factor is the ratio of the undercut d to the depth of etch T. This ratio must be considered when

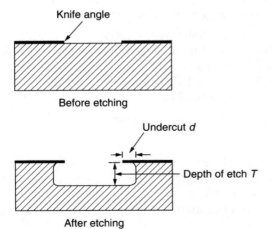

Figure 3.2 Etch factor after CHM.

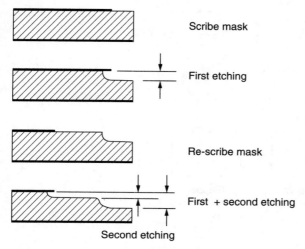

Figure 3.3 Contour cuts by CHM.

scribing the mask using templates. A typical etch factor of 1:1 occurs at a cut depth of 1.27 mm. Deeper cuts can reduce this ratio to 1:3. The radii of the fillet produced will be approximately equal to the depth of etch. For simultaneous machining of multiple parts, racks or handling fixtures are frequently used to facilitate the submersion of the work in the chemical reagent and for subsequent rinsing. After rinsing the chemicals from the workpiece, the demasking is accomplished by hand stripping, mechanical brushing, or chemical stripping. Some chemicals leave a film of smut on the machined surface, which can be removed by other chemicals or frequently by brushing.

CHM will not eliminate surface irregularities, dents, scratches, or waviness. Successive steps of mask removal and immersion as shown in Fig. 3.3 can achieve stepped cuts. Tapered cuts (Fig. 3.4), can also be produced without masking the workpiece by controlling the depth and rate of immersion or withdrawal and the number of immersions. Continuous tapers, as great as 0.060 mm/mm for aluminum and 0.010 mm/mm for steel alloys, have been machined on a production basis (*Metals Handbook*, 1989).

3.1.2 Tooling for CHM

Tooling for CHM is relatively inexpensive and simple to modify. Four different types of tools are required: maskants, etchants, scribing templates, and accessories.

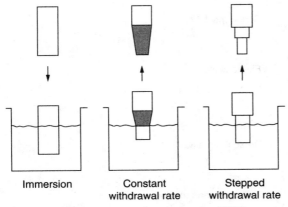

Figure 3.4 Machining tapers and steps by CHM.

3.1.2.1 Maskants. Maskants are generally used to protect parts of the workpiece where CD action is not needed. Synthetic or rubber base materials are frequently used. Table 3.1 shows the different maskants and etchants for several materials together with the etch rate and etch factor. Maskants should, however, possess the following properties:

1. Be tough enough to withstand handling
2. Adhere well to the workpiece surface
3. Scribe easily
4. Be inert to the chemical reagent used

TABLE 3.1 Maskants and Etchants for Different Workpiece Materials

Workpiece	Etchant	Maskant	Etch rate, mm/min	Etch factor
Aluminum	$FeCl_3$	Polymers	0.013–0.025	1.5–2.0
	NaOH	Polymers	0.020–0.030	
Magnesium	HNO_3	Polymers	1.0–2.0	1.0
Copper	$FeCl_3$	Polymers	2.0	2.5–3.0
	$CuCl_3$		1.2	
Steel	$HCl:HNO_3$	Polymers	0.025	2.0
	$FeCl_3$		0.025	
Titanium	HF	Polymers	0.025	1.0
	$HF:HNO_3$			
Nickel	$FeCl_3$	Polyethylene	0.13–0.038	1.0–3.0
Silicon	$HNO_3:HF:H_2O$	Polymers	Very slow	

SOURCE: Tlusty (1999).

5. Be able to withstand the heat generated by etching
6. Be removed easily and inexpensively after etching

Multiple coats of maskant are frequently used to increase the etchant resistance and avoid the formation of pinholes on the machined surfaces. When thicker, rougher dip or spray coatings are used, deeper cuts that require long exposure time to the etchant can be achieved. Dip, brush, spray, roller, and electrocoating as well as adhesive tapes can be used to apply masks.

Spraying the mask on the workpiece through silk screen, on which the desired design is imposed, combines the maskant application with the scribing operation since no peeling is required. The product quality is, therefore, improved as is the ability to generate finer details. However, the thin coating layer applied when using silk screens will not resist etching for a long time as will the cut-and-peel method. Photoresist masks, which are used in photochemical milling (PCM), also combine both the coating and scribing operations. The relatively thin coats applied as dip or spray coats will not withstand rough handling or long exposure times to the etchant. However, photoresist masks ensure high accuracy, ease of repetition for multiple-part etching, and ease of modification. The accuracy obtained for lateral dimensions depends on the complexity of the masking. Typical tolerances for the different masks are as follows:

- Cut-and-peel masks   ±0.179 mm
- Silk-screen resist   ±0.077 mm
- Photoresist   ±0.013 mm

3.1.2.2 Etchants. Etchants (see Table 3.1) are acid or alkaline solutions maintained within a controlled range of chemical composition and temperature. Their main technical goals are to achieve the following:

1. Good surface finish
2. Uniformity of metal removal
3. Control of selective and intergranular attack
4. Control of hydrogen absorption in the case of titanium alloys
5. Maintenance of personal safety
6. Best price and reliability for the materials to be used in the construction of the process tank
7. Maintainance of air quality and avoidance of possible environmental problems

8. Low cost per unit weight dissolved
9. Ability to regenerate the etchant solution and/or readily neutralize and dispose of its waste products

3.1.2.3 Scribing templates. Scribing templates are used to define the areas for exposure to the chemical machining action. The most common workpiece scribing method is to cut the mask with a sharp knife followed by careful peeling of the mask from the selected areas. Layout lines or simple templates of metal or fiberglass guide the scribing process. The etch factor allowance must be included in any method used for the scribing operation. The negative (used in PCM) or its layout and the template or the silk screen must allow for the degree of undercutting expected during etching. Figure 3.5 shows numerical control (NC) laser scribing of masks for CHM of a large surface area.

3.1.2.4 Accessories. Accessories include tanks, hooks, brackets, racks, and fixtures. These are used for single- or multiple-piece handling into and out of the etchants and rinses.

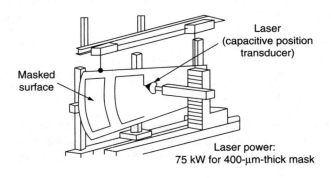

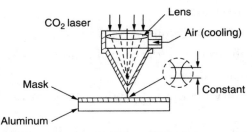

Figure 3.5 Laser cutting of masks for CHM of large surfaces (Tlusty, 1999).

3.1.3 Process parameters

CHM process parameters include the reagent solution type, concentration, properties, mixing, operating temperature, and circulation. The process is also affected by the maskant and its application. These parameters will have direct impacts on the workpiece regarding the following:

1. Etch factor (d/T)
2. Etching and machining rate
3. Production tolerance
4. Surface finish

To machine high-quality and low-cost parts using CHM, we must consider the heat treatment state of the workpiece, the grain size and range of the workpiece material, the size and finish control prior to CHM, the direction of rolling and weld joints, and the degree of cold work.

3.1.4 Material removal rate

The material removal or etch rate depends upon the chemical and metallurgical uniformity of the workpiece and the uniformity of the solution temperature. As shown in Figs. 3.6 and 3.7, castings, having the largest grain size, show the roughest surface together with the lowest

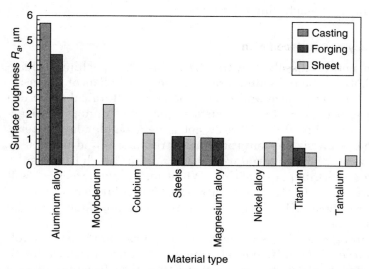

Figure 3.6 CHM average roughness of some alloys after removing 0.25 to 0.4 mm (El-Hofy, 1995).

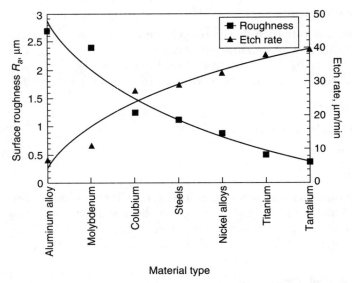

Figure 3.7 Surface roughness and etch rate of some alloys after removing 0.25 to 0.4 mm (El-Hofy, 1995).

machining rate. Rolled metal sheets have the highest machining rate accompanied by the best surface quality. Etching rates were high for hard metals and were low for softer ones (*Metals Handbook*, 1989). Generally, the high etch rate is accompanied by a low surface roughness and, hence, narrow machining tolerances.

3.1.5 Accuracy and surface finish

In CHM, the metal is dissolved by the CD action. This machining phase takes place both at the individual grain surfaces as well as at the grain boundaries. Fine grain size and homogenous metallurgical structure are, therefore, necessary, for fine surface quality of uniform appearance. Surfaces machined by CHM do not have a regular lay pattern. Based on the grain size, orientation, heat treatment, and previously induced stresses, every material has a basic surface finish that results from CHM for a certain period of time. While surface imperfections will not be eliminated by CHM, any prior surface irregularities, waviness, dents, or scratches will be slightly altered and reproduced in the machined surface.

The machining rate affects the surface roughness and hence the tolerance produced. Generally, slow etching will produce a surface finish similar to the original one. Figure 3.7 shows typical surface roughnesses for different materials. The orientation of the areas being etched with respect to the rolling direction or the direction of the grain in the workpiece is also important for good CHM surfaces. The depth of cut tolerance increases when machining larger depths at high machining rates. Aluminum and magnesium alloys can be controlled more closely than steel, nickel, or titanium alloys. An etching rate of 0.025 mm/mm with tolerances of ±10 percent of the cut width can be achieved depending on the workpiece material and depth of cut.

The surface roughness is also influenced by the initial workpiece roughness. It increases as the metal ion concentration rises in the etchant. For low machining depths, <200 µm, the roughness sharply increases with the depth of cut, while at higher depths a slight change in the roughness is evident. Figure 3.7 shows the dependence of the surface roughness and etch rate on the workpiece material. Typically, surface roughnesses of 0.1 to 0.8 µm, depending on the initial roughness, can be obtained. However, under special conditions, roughnesses of 0.025 to 0.05 µm become possible (*Machining Data Handbook*, 1997).

CHM can affect the mechanical properties of the machined parts when the surface layers have different mechanical properties from those of the base metal. The removal of such layers results in a change in the average mechanical properties of the finished parts. In this regard surface conditions such as a titanium oxide layer (alpha case), decarburized layer, and recast structure are easily removed by CHM, resulting in an improvement in the properties of the finished parts. Some loss of fatigue properties has been reported after CHM of aluminum; however, shot peening or grit blasting can restore it.

3.1.6 Advantages

The process has the following advantages:

- Weight reduction is possible on complex contours that are difficult to machine using conventional methods.
- Simultaneous material removal, from all surfaces, improves productivity and reduces wrapping.
- No burrs are formed.
- No stress is introduced to the workpiece, which minimizes the part distortion and makes machining of delicate parts possible.
- A continuous taper on contoured sections is achievable.
- The capital cost of equipment, used for machining large components, is relatively low.
- Design changes can be implemented quickly.
- A less skilled operator is needed.

- Tooling costs are minor.
- The good surface quality in addition to the absence of burrs eliminates the need for finishing operations.
- Multiple parts having fine details can be machined by the gang method.
- Decorative finishes and extensive thin-web areas are possible.
- There are low scrap rates (3 percent).

3.1.7 Limitations

CHM does have limitations and areas of disadvantage:

- Only shallow cuts are practical: up to 12.27 mm for sheets and plates, 3.83 mm on extrusions, and 6.39 mm on forgings.
- Handling and disposal of chemicals can be troublesome.
- Hand masking, scribing, and stripping can be time-consuming, repetitive, and tedious.
- Surface imperfections are reproduced in the machined parts.
- Metallurgical homogeneous surfaces are required for best results.
- Deep narrow cuts are difficult to produce.
- Fillet radii are fixed by the depth of cut.
- Porous castings yield uneven etched surfaces.
- Welded areas frequently etch at rates that differ from the base metal.
- Material removal from one side of residually stressed material can result in a considerable distortion.
- The absence of residual stresses on the chemically machined surfaces can produce unfavorable fatigue strength compared with the processes that induce compressive residual stresses.
- Hydrogen pickup and intergranular attack are a problem with some materials.
- The straightness of the walls is subject to fillet and undercutting limitations.
- Scribing accuracy is limited and complex designs become expensive.
- Steep tapers are not practical.

3.1.8 Applications

All the common metals including aluminum, copper, zinc, steel, lead, and nickel can be chemically machined. Many exotic metals such as titanium, molybdenum, and zirconium, as well as nonmetallic materials including glass, ceramics, and some plastics, can also be used with the process. CHM applications range from large aluminum airplane wing parts to minute integrated circuit chips. The practical depth of cut ranges between 2.54 to 12.27 mm. Shallow cuts in large thin sheets are of the most popular application especially for weight reduction of aerospace components. Multiple designs can be machined from the same sheet at the same time. CHM is used to thin out walls, webs, and ribs of parts that have been produced by forging, casting, or sheet metal forming, as shown in Fig. 3.8. Further process applications related to improving surface characteristics include the following:

1. Elimination of alpha case from titanium forgings and superplastic-formed parts
2. Elimination of the decarburized layer from low alloy steel forgings
3. Elimination of the recast layer from parts machined by EDM
4. Removal of sharp burrs from conventionally machined parts of complex shapes
5. Removal of a thin surface from forgings and castings prior to penetration inspection below the surface (required for the detection of hidden defects)

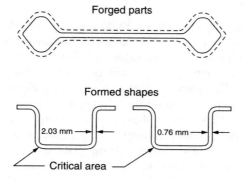

Figure 3.8 Thinning of parts by CHM (Tlusty, 1999).

3.2 Photochemical Milling

3.2.1 Introduction

Photochemical milling (PCM) is a variation of chemical milling (CHM) where the chemically resistant mask is applied to the workpiece by photographic techniques. The two processes are quite similar because they both use chemicals to remove metal by the CD action and some of the steps required in both cases are similar. CHM is usually used on three-dimensional parts originally formed by another manufacturing process, such as forging and casting of irregular shapes. As with photochemical machining, areas not to be machined are masked from the CD action of the chemical solution. Photochemical machining, therefore, creates new parts from thin materials, rather than simply smoothing or altering parts formed by other manufacturing methods. The process is sometimes called photochemical blanking (PCB) since flat, thin gauge complex parts can be machined to high precision in thicknesses ranging from 0.013 to 1.5 mm. Sometimes photochemical machining is used to surface-etch components with lettering or graphics where the etchant works its way to only a certain depth in the material. As in the case of chemical milling, the depth of etch is controlled by the time a component is immersed in a particular etchant. Neither photochemical machining nor chemical milling should be confused with photo forming (PF), which is the process of electroplating metals over a mandrel. The photo formed components are removed from the mandrel after the proper thickness of material has been deposited.

3.2.2 Process description

Figure 3.9 shows the flow diagram of the PCM process. The first step includes the production of the required shape on a photographic film or glass plate, termed the photo-tool. Computer-aided design (CAD) artwork creates the required part shape, which is used as the primary image for the photo-tool. The sheet metal is chemically cleaned, and coated with a light-sensitive photoresist film. The photoresist will adhere to the part surface and act as a stencil resist protecting the surface during etching. In some cases, the photoresist is a liquid and the part has to be dip coated and dried. Most often, photo-tools are used in precisely registered pairs—one on the top, one on the bottom, with the material to be machined sandwiched in between. This permits the material to be etched from both sides, thus minimizing undercutting of the photoresist and producing straighter sidewalls.

The resist laminated, or coated, metal is then placed under the photo-tool and exposed, in a vacuum, to an ultraviolet light source. This transfers the image precisely onto the resist and, after developing, becomes

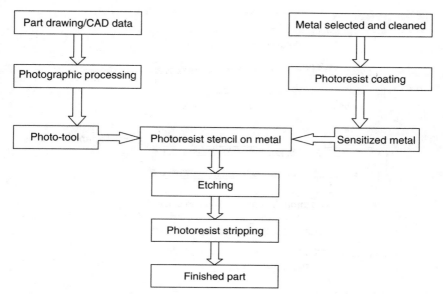

Figure 3.9 PCM flow diagram.

a replica of the desirable geometry. The exposed image is then developed by immersion or spraying. Each photoresist has its own developing solution, such as water, alkaline solution, hydrocarbons, or solvents. The exposed material is then washed to remove the unexposed photoresist on the areas to be chemically etched as illustrated by Allen (1993).

The imaged metal passes through the acid etch spray or dip where it is selectively dissolved away. Various etchants are available for different materials. The right choice of etchant depends on issues such as cost, quality, depth of etch, and rate of material removal. Figure 3.10 shows the main steps of photochemical machining. After applying etchants by spraying or immersion, the parts are rinsed and dried. Protective resist is removed from machined parts with chemicals or with mechanical techniques along with chemicals.

3.2.3 Applications

Aluminum, copper, zinc, steel, lead, nickel, titanium, molybdenum, zirconium, glass, ceramics, and some plastics are photochemically machined. Very high tempered or brittle materials are excellent candidates for photochemical machining because traditional machining causes breakage or stress-concentration points. The process also works well on springy materials, which are difficult to punch. PCM lends itself for

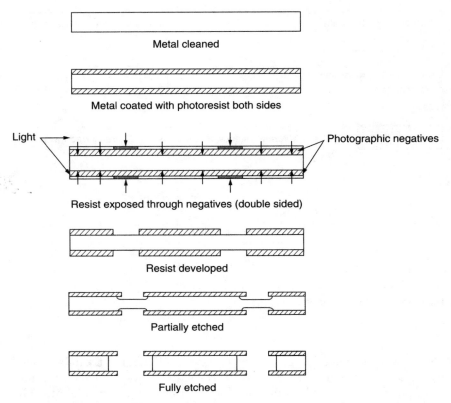

Figure 3.10 PCM steps (Tlusty, 1999).

decorative and graphics industries where signs and labels are produced as shown in Fig. 3.11. Materials undergoing PCM must be thin, with a thickness between 0.013 and 1.5 mm. The materials must also be flat so that they can later be bent to shape and assembled into other components. In this regard, etching fold lines to flat components for fabrication of boxes and enclosures is a typical PCM application. Products made by photochemical machining are generally found in the electronic, automotive, aerospace, telecommunication, computer, medical, and other industries. Typical components (Fig. 3.12), include filters and screens, gaskets, lead frames, contacts, connectors, probes, and flat springs. Figure 3.13 shows different photochemically machined patterns.

3.2.4 Advantages

In addition to the general advantages of CHM, PCM ensures the following merits:

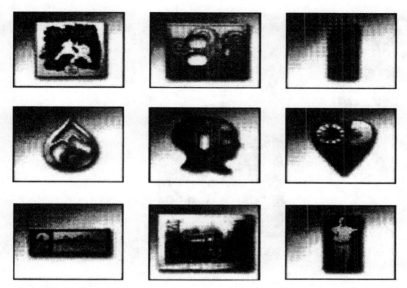

Figure 3.11 PCM of decorative and artistic designs (www.aculineetch.com/).

- A relatively low cost per unit, especially at low production volumes of complex designs because the tooling used is very inexpensive compared to shearing punches and dies.
- Lead times are often small compared to that required by processes that require hard tooling.

Figure 3.12 Typical PCM blanks (www.chemart.com/).

Figure 3.13 Samples of PCM machined patterns (www.greatlakesetching.com/).

- Some design changes require simple alterations to the photochemical machining process, such as the time of etch or the type of etchant, which change features such as the hole size and depth of etch.
- Final parts are produced in the same manner as the prototypes.
- The process is burr-free.
- It does not change the hardness, grain structure, or ductility of metals, while metal shearing imparts stresses in the components and laser machining creates a heat-affected zone.
- Because tooling is made by photographic techniques, patterns can be reproduced easily.

3.3 Electropolishing

3.3.1 Introduction

Mechanical polishing entails using abrasive particles adhered to the resilient wheels of wood, felt, leather, canvas, or fabric to produce smooth surfaces. The process is used to impart a high-grade finish to a surface for the sake of good appearance. However, mechanical polishing leaves a layer of disturbed structure since the surface does not have the same properties of the bulk metal. The mechanically polished surface yields an abundance of scratches, strains, metal debris, and embedded abrasives, which reduce the mechanical strength. Further finishing by lapping or buffing, while decreasing the surface roughness, never completely removes the debris and the damaged layer caused by the previous mechanical polishing. The drawbacks of such a conventional polishing method can be overcome using the nonconventional electropolishing (EP). This process is basically the opposite of electroplating where the part to be finished is the anode instead of the cathode. EP dates back to 1935 when copper and zinc were successfully electropolished by the Germans.

Electropolishing is a diffusion-controlled process, which takes place at the limiting current of the anodic dissolution of the metal. Figure 3.14 shows the relationship between the current density and the anode potential. Accordingly, a matt surface occurs between points A and B while polishing is achieved between B and C. Along CD the polishing action is, normally, accompanied by surface pitting due to the rupture of the anodic layer by gas evolution (McGeough, 1974).

Figure 3.15 shows the main components of a typical polishing cell. A direct current is accordingly introduced into the part, which is hung from the central electrode and is surrounded by cathodes that are negatively charged. The electropolishing medium is a liquid mixture of several acids and insoluble salts.

During electropolishing the anodic surface is subjected to the combined effects of oxygen evolution that occurs with electrochemical metal removal and saturation of the surface with dissolved metal. A film of varying thickness is normally formed over the microdepressions and is

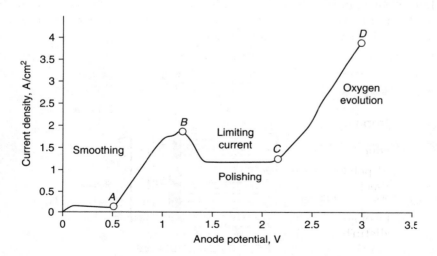

Figure 3.14 Current density and anode potential during the EP process.

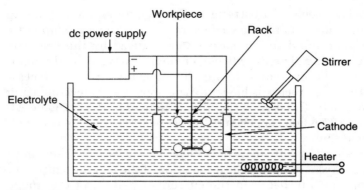

Figure 3.15 Electropolishing schematic.

thinnest over microsurface projections. The electrical resistance is, therefore, minimum at the peaks of the surface irregularities resulting in the highest rate of anodic dissolution process (Fig. 3.16). Under such conditions, electropolishing selectively removes microscopic peaks faster than the rate of attack on the corresponding valleys. The basic surface of the metal becomes bright, clean, and microscopically smooth. The produced surface possesses the true properties of the bulk metal and the true crystalline structure of the metal without the distortion produced by the cold working that always accompanies mechanical finishing processes. Furthermore, no fine directional lines from abrasive polishing are left at the EP surface.

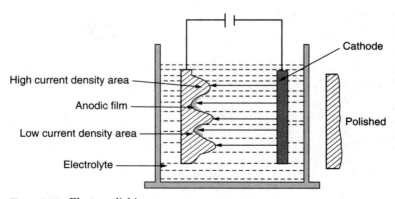

Figure 3.16 Electropolishing process.

3.3.2 Process parameters

EP is affected by many parameters that have a direct impact on the produced surface quality and process productivity. These include the following:

1. Workpiece material and condition
2. Original surface roughness
3. Current density
4. Applied voltage
5. Acid type, temperature, and agitation

During EP it is recommended that you use a proper electrolyte, maintain its temperature and chemical composition, and supply a ripple-free DC power at the correct voltage. Table 3.2 presents suitable electrolyte conditions and current densities for electropolishing of different metals and alloys.

3.3.3 Applications

For economical and high-quality electropolishing, cold rolled metals are the most important materials. Cast items usually maintain the texture of the sand mold. They have coarse grains and include alloying elements (phosphorus, lead, and silicon) that improve metal fluidity. Such a coarse texture is not removed by electropolishing; however, the surface can be made brilliant. In general, coarse grain structures and phosphorus-, lead-, and silicon-rich phases produce rough surfaces.

TABLE 3.2 Electrolytes and Current Densities for EP of Different Metals

Metal	Electrolyte	Current density, A/cm^2
Carbon steel	182 mL $HCLO_4$, 765 mL acetic anhydride, 50 mL H_2O	0.04–0.06
Stainless steel	60% H_3PO_4, 20% H_2SO_4, 20% H_2O	1.8 at 80°C
	37% H_3PO_4, 56% glycerine, 7% H_2O	1.0 at 120°C
Copper	H_3PO_4	0.04
Alpha brass	$H_3PO_4Cr_{11}$, $NaCr_2O_2$, H_2SO_4, HF, Cr_2H_2, COOH	0.14–0.35 at 50°C
Phosphor bronze	33% HNO_3	0.75–1.5
Cupronickel	67% CH_2OH	
Nickel	75% H_2SO_4	0.25
Aluminum	$NCIO_3$	0.04–0.08
Magnesium	37% H_2PO_4, 63% C_2H_3OH	0.005

The EP process finds many applications which have been reported by *www.globalstainlesstech.com/* and Brown (1998):

1. Preparing surfaces for electroplating, which creates an opportunity for the best possible adherence.
2. Producing the ultimate finish for clean-room tables, chairs, and waste containers.
3. Polishing light fixtures of electrical conduits; outlet boxes; and medical, surgical, and food processing equipment.
4. Deburring and breaking sharp edges resulted from hand filing and honing and grinding of cutting tools.
5. Metallographic examination.
6. Removing scale or distortions caused by annealing, nitriding, carburizing, welding, or soldering.
7. Removing skin that remains on metals after casting or forging, which allows for subsequent machining operations with less effort, time, and tool wear.
8. Removing hardened and stressed surface layers, which improve the surface life of a part.
9. Improving adhesion for coatings such as paint and plasma spraying.
10. Smoothing of the surface to increase reflectivity and thus creating a bright appearance.
11. Removing burrs, occlusions, and other metalworking marks, which makes it easier to clean the surface and avoid microbiological contaminants.
12. Micromachining of metals and alloys.
13. Polishing a large number of parts simultaneously.

3.3.4 Process limitations

- The process cannot smear over and cover up defects such as seams and nonmetallic inclusions in the metal.
- Multiphase alloys in which one phase is relatively resistant to anodic dissolution are usually not amenable to electropolishing.
- Rough scratches are not removed even by a considerable amount of electropolishing.
- Metals containing a high percentage of silicon, lead, and sulfur can be troublesome.

- Electropolishing is more suitable for removing small scratches and imperfections than for smoothing out any type of surface waviness.
- Coatings, in contrast to wrought metals, will not polish to a brightness or smoothness.
- The base metal condition affects the electropolishing.
- Nonmetallic inclusions, improper annealing, overpickling, heat scale, large grain size, directional roll marks, and improper cold reduction leads to poor electropolished surfaces.

References

Allen, D. (1993). "Progress Towards Clean Technology for Photochemical Machining," *Annals of CIRP*, 42(1):179–200.
Bellows, G. (1977). "Chemical Machining, Production with Chemistry," MDC 77-102, Machinability Data Centre (MDC) Machining Process Series Metcut Research Associates Inc., Cincinnati, OH.
Brown, J. (1998). *Advanced Machining Technology Handbook*. New York: McGraw-Hill.
El-Hofy, H. (1995). "Machinability Indices of Some Non-conventional Machining Processes," *Alexandria Engineering Journal* (AEJ), 34(3):231–245.
El-Hofy, H. (1996). "Surface Generation in Non-conventional Machining," *Sixth MDP Conf.*, Cairo, pp. 203–213.
Machining Data Handbook (1997). 3rd ed., Vol. 2, Cincinnati, OH: Machinability Data Center, Institute of Advanced Manufacturing.
McGeough, J. A. (1974). *Principles of Electrochemical Machining*. London, New York: Chapman and Hall.
Metals Handbook (1989). 9th ed., Vol. 16, *Machining*. Meterials Park, OH: ASM International.
Tlusty, G. (1999). *Manufacturing Processes and Equipment*. Upper Saddle River, NJ: Prentice-Hall.
www.acctechpm.com/
www.aculineetch.com/etchingep.html
www.chemart.com/
www.globalstainlesstech.com/Electropolishing/Diagram1/diagram1.html
www.greatlakesetching.com/patterns.html
www.Lkengineering.com/index.html
www.photo-sciences.comw:1000/start.html

Chapter 4

Electrochemical Processes

4.1 Electrochemical Machining

4.1.1 Introduction

Electrochemical machining (ECM) is a modern machining process that relies on the removal of workpiece atoms by electrochemical dissolution (ECD) in accordance with the principles of Faraday (1833). Gusseff introduced the first patent on ECM in 1929, and the first significant development occurred in the 1950s, when the process was used for machining high-strength and heat-resistant alloys.

4.1.2 Principles of electrolysis

Electrolysis occurs when an electric current passes between two electrodes dipped into an electrolyte solution. The system of the electrodes and the electrolyte is referred to as the *electrolytic cell*. The chemical reactions, which occur at the electrodes, are called the anodic or cathodic reactions. ED of the anodic workpiece forms the basis for ECM of metals. The amount of metal dissolved (removed by machining) or deposited is calculated from Faraday's laws of electrolysis, which state that

1. The amount of mass dissolved (removed by machining), m, is directly proportional to the amount of electricity.

$$m \propto It$$

2. The amount of different substances dissolved, m, by the same quantity of electricity (It) is proportional to the substances' chemical equivalent weight ε.

$$m \propto \varepsilon$$

and

$$\varepsilon = \frac{A}{Z}$$

where I = electrolyzing current, A
t = machining time, min
ε = chemical equivalent weight, g
A = atomic weight
Z = workpiece valence

4.1.3 Theory of ECM

ECM uses a direct current at a high density of 0.5 to 5 A/mm^2 and a low voltage of 10 to 30 V. The machining current passes through the electrolytic solution that fills the gap between an anodic workpiece and a preshaped cathodic tool. The electrolyte is forced to flow through the interelectrode gap at high velocity, usually more than 5 m/s, to intensify the mass and charge transfer through the sublayer near the anode. The electrolyte removes the dissolution products, such as metal hydroxides, heat, and gas bubbles, generated in the interelectrode gap. McGeough (1988) claimed that when a potential difference is applied across the electrodes, several possible reactions occur at the anode and the cathode. Figure 4.1 illustrates the dissolution reaction of iron in a sodium chloride (NaCl) water solution as an electrolyte. The result of electrolyte dissociation and NaCl dissolution leads to

$$H_2O \rightarrow H^+ + OH^-$$
$$NaCl \rightarrow Na^+ + Cl^-$$

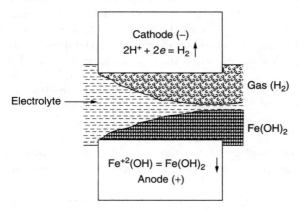

Figure 4.1 Electrochemical reactions during ECM of iron.

The negatively charged anions OH⁻ and Cl⁻ move toward the anode, and the positively charged cations of H^+ and Na^+ are directed to the cathode. At the anode, Fe changes to Fe^{++} by losing two electrons.

$$Fe \rightarrow Fe^{++} + 2e$$

At the cathode, the reaction involves the generation of hydrogen gas and the hydroxyl ions.

$$2H_2O + 2e \rightarrow H_2 + 2(OH)^-$$

The outcome of these electrochemical reactions is that iron ions combine with other ones to precipitate out as iron hydroxide, $Fe(OH)_2$.

$$Fe + 2H_2O \rightarrow Fe(OH)_2 + H_2$$

The ferrous hydroxide may react further with water and oxygen to form ferric hydroxide, $Fe(OH)_3$.

$$4Fe(OH)_2 + 2H_2O + O_2 \rightarrow 4Fe(OH)_3$$

With this metal-electrolyte combination, electrolysis has involved the dissolution of iron, from the anode, and the generation of hydrogen, at the cathode (McGeough, 1974).

4.1.4 ECM equipment

Figure 4.2 shows the main components of the ECM machine: the feed control system, electrolyte supply system, power supply unit, and workpiece holding device. As shown in Fig. 4.3, the feed control system is responsible for feeding the tool at a constant rate during equilibrium machining. The power supply drives the machining current at a constant dc (continuous or pulsed) voltage. The electrolyte-feeding unit supplies the electrolyte solution at a given rate, pressure, and temperature. Facilities for electrolyte filtration, temperature control, and sludge removal are also included. ECM machines are capable of performing a wide range of operations such as duplicating, sinking, and drilling. Semiautomatic and fully automated facilities are used for large-size machining, such as deburring in the automotive industry. ECM machines, in contrast to conventional machine tools, are designed to stand up to corrosion attack by using nonmetallic materials. For high strength or rigidity, metals with nonmetallic coatings are recommended at *www.unl.edu/*.

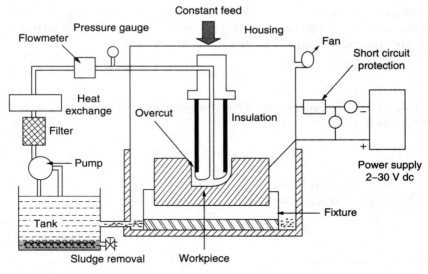

Figure 4.2 ECM elements.

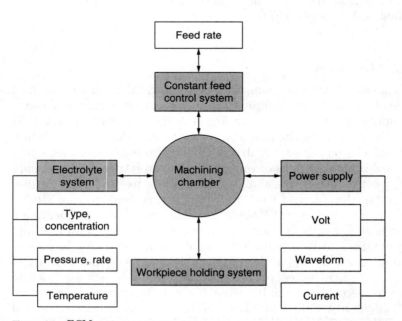

Figure 4.3 ECM system components.

4.1.4.1 Power supply.
The dc power supply for ECM has the following features:

1. Voltage of 2 to 30 volts (V) (pulsed or continuous)
2. Current ranges from 50 to 10,000 amperes (A), which allow current densities of 5 to 500 A/cm^2
3. Continuous adjustment of the gap voltage
4. Control of the machining current in case of emergency
5. Short circuit protection in a matter of 0.001 s
6. High power factor, high efficiency, small size and weight, and low cost

4.1.4.2 Electrolytes.
The main functions of the electrolytes in ECM are to

1. Create conditions for anodic dissolution of workpiece material
2. Conduct the machining current
3. Remove the debris of the electrochemical reactions from the gap
4. Carry away the heat generated by the machining process
5. Maintain a constant temperature in the machining region

The electrolyte solution should, therefore, be able to (*www.unl.edu/*)

1. Ensure a uniform and high-speed anodic dissolution
2. Avoid the formation of a passive film on the anodic surface (electrolytes containing anions of Cl, SO_4, NO_3, ClO_3, and OH are often recommended)
3. Not deposit on the cathode surface, so that the cathode shape remains unchanged (potassium and sodium electrolytes are used)
4. Have a high electrical conductivity and low viscosity to reduce the power loss due to electrolyte resistance and heat generation and to ensure good flow conditions in the extremely narrow interelectrode gap
5. Be safe, nontoxic, and less erosive to the machine body
6. Maintain its stable ingredients and pH value, during the machining period
7. Have small variation in its conductivity and viscosity due to temperature rise
8. Be inexpensive and easily available

The most common electrolytes used are sodium chloride (NaCl), sodium nitrate ($NaNO_3$), and, sodium hydroxide. Industrial ECM operations usually involve using mixed electrolytes to meet multiple requirements

TABLE 4.1 Electrolytes and Machining Rates for ECM of Various Work Materials

Work material	Electrolyte Constituent	Concentration g/L H_2O	Removal rate, $mm^3/(min \cdot A)$
Grey iron	NaCl	300	2.0
	$NaNO_3$	600	2.0
White cast iron	$NaNO_3$	600	1.6
Steel; hardened tool steel	$NaClO_3$	780	2.0
Steel; iron-, and nickel-,	$NaNO_3$	600	2.1
and cobalt-base alloys	NaCl or KCl	300	2.1
Copper & copper alloys	NaCl or KCl	300	4.4
	$NaNO_3$	600	3.3
Tungsten	NaOH	180	1.0
Titanium alloys	NaCl or KCl	120	1.6
Molybdenum	NaOH	180	1.0
	NaCl or KCl	300	1.0
Zirconium	NaCl or KCl	300	2.1

SOURCE: Data from *Metals Handbook* (1989).

as shown in Table 4.1. The selection of the ECM electrolyte depends on the workpiece material, the desired dimensional tolerance, the surface finish required, and the machining productivity. During ECM, the electrolyte plays an important role in dimensional control. As shown in Fig. 4.4, sodium nitrate solution is preferable, because the local metal removal rate is high at the small gap locations where both the current density and the current efficiency are high. Additionally, the local removal rate is low at the larger gap locations where both the current density and current efficiency are low. This results in the gap distribution tending toward uniformity.

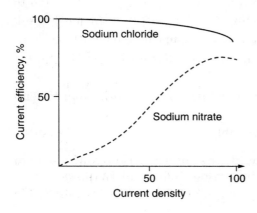

Figure 4.4 Effect of current density on current efficiency for different electrolytes.

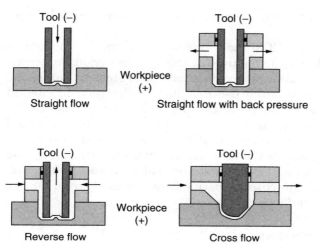

Figure 4.5 Modes of electrolyte feeding in ECM.

The current efficiency in ECM depends on the anodic material and the electrolyte. When the pulsed voltage is applied instead of the commonly used continuous voltage, proper use of pulse parameters (e.g., pulse on-times) can significantly improve the current efficiency and surface quality. Depending on the tool shape and type of the machining operation, several methods of supplying electrolyte to the machining gap are shown in Fig. 4.5. The choice of the electrolyte supply method depends on the part geometry, machining method, required accuracy, and surface finish. Typical electrolyte conditions include a temperature of 22 to 45°C, a pressure between 100 to 200 kPa, and a velocity of 25 to 50 m/s.

4.1.4.3 Tools. The design of a suitable tool for a desired workpiece shape, and dimension forms a major problem. The workpiece shape is expected to be greater than the tool size by an oversize. In determining the geometry of the tool to be used under steady-state conditions, many variables should be specified in advance such as the required shape of the surface to be machined, tool feed rate, gap voltage, electrochemical machinability of the work material, electrolyte conductivity, and anodic and cathodic polarization voltages. With computer integrated manufacturing, cathodes are produced at a lower cost and greater accuracy. Computer-aided design (CAD) systems are used first to design a cathodic tool. This design is programmed for CNC production by milling and turning. After ECM, the coordinate measuring machine inspects the workpiece produced by this tool and sends data back to the CAD computer-aided manufacturing (CAM) unit for analysis of the results.

Iterations of the cathodic tool are made so that the optimum tool design is selected.

The material used for ECM tools should be electrically conductive and easily machinable to the required geometry. The various materials used for this purpose include copper, brass, stainless steel, titanium, and copper tungsten. Tool insulation controls the side electrolyzing current and hence the amount of oversize. Spraying or dipping is generally the simplest method of applying insulation. Durable insulation should ensure a high electrical receptivity, uniformity, smoothness, heat resistance, chemical resistance to the electrolyte, low water absorption, and resistance against wear in the machine guides and fixtures. Teflon, urethane, phenol, epoxy, and powder coatings are commonly used for tool insulation (*Metals Handbook*, 1989).

4.1.5 Basic working principles

The simplest case to consider is that of plane-parallel electrodes normal to the feed direction as described by Tipton (1971) and shown in Fig. 4.6. Consider an electrolyte of conductivity κ and density ρ_e that flows at a mean velocity u, in the direction of increasing x, in a channel. The channel is assumed to extend to the left of the origin $x = 0$ where the tool and workpiece start, so that the flow has reached a steady state and the inlet conditions can be neglected. All properties of the system are assumed to be independent of the z direction. The position of the workpiece surface relative to the tool and hence the gap thickness is represented by the coordinate y. The workpiece surface moves away from the tool surface in the direction of increasing y at a rate proportional to the current density J and equal to

$$\frac{\gamma \varepsilon J}{F \rho} = \frac{\gamma \varepsilon (v - \Delta v) \kappa}{F \rho y}$$

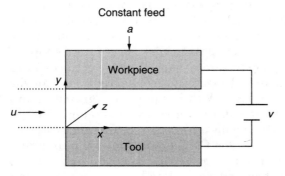

Figure 4.6 Working gap with plane-parallel electrodes at a constant gap voltage.

At the feed rate a, in the direction of decreasing y, the workpiece rate of change of position dy/dt can be written as

$$\frac{dy}{dt} = \frac{\gamma\varepsilon(v-\Delta v)\kappa}{F\rho y} - a$$

where γ = current efficiency of the dissolution process, %
F = Faraday's constant, 96,500 C/g per ion
κ = electrolyte conductivity, $\Omega^{-1}\text{mm}^{-1}$
ρ = density of anode material, g/mm^3
ε = chemical equivalent weight
v = applied voltage, V
Δv = overvoltage, V

The current efficiency γ is defined as the ratio of the observed amount of metal dissolved to the theoretical amount predicted from Faraday's laws for the same specified conditions of electrochemical equivalence, current, etc. Apparent current efficiency values may be due to

1. The choice of wrong valence
2. Passivation of the anodic surface
3. Grain boundary attack, which causes the removal of metal grains by electrolyte forces
4. Gas evolution at the anode surface

It is convenient to write the machining constant C for the particular workpiece-electrolyte combination (m$^2 \cdot$ min^{-1}) as

$$C = \frac{\gamma\varepsilon(v-\Delta v)\kappa}{F\rho}$$

Then,

$$\frac{dy}{dt} = \frac{C}{y} - a$$

Integrating

$$\int_0^t dt = \int_{y_0}^y \frac{y}{C-ay}\,dy$$

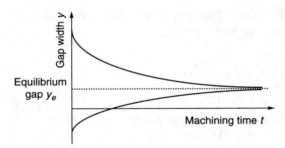

Figure 4.7 Attainment of equilibrium gap at a constant feed rate and various initial gaps.

$$t = \int_{y_0}^{y} \frac{C}{y} \frac{1}{C - ay} \, dy$$

$$= \frac{1}{a}(y - y_0) + \frac{C}{a^2} \ln\left(\frac{C - ay_0}{C - ay}\right)$$

Figure 4.7 shows how y varies with time approaching equilibrium value y_e asymptotically. For the case of zero feed ($a = 0$),

$$\frac{dy}{dt} = \frac{C}{y}$$

If the initial position of the workpiece surface at $t = 0$ is y_0, then

$$y = \sqrt{y_0^2 + 2Ct}$$

Therefore, the gap increases in proportion to the square root of time as shown in Fig. 4.8. At a constant feed rate a, the workpiece surface will

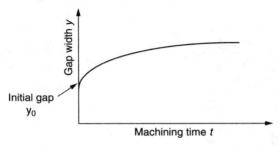

Figure 4.8 Variation of initial gap with machining time at zero feed.

be stationary and hence the gap thickness becomes constant, when

$$\frac{dy}{dt} = 0$$

or

$$y_e = \frac{C}{a}$$

Under such conditions, the metal removal rate per unit area (C/y_e) is equal to the feed rate (a) and the corresponding gap thickness is called the equilibrium gap y_e. As shown in Fig. 4.7, if the gap thickness is greater than y_e, the metal removal rate is less than the feed rate, so the gap closes up toward y_e. If the initial gap is less than y_e, the removal rate is greater than the feed rate, so the gap always tends toward the equilibrium value y_e as the process proceeds. During ECM drilling, decreasing the feed rate leads to wider machining gaps and hence a lack of dimensional accuracy. On the other hand, too small of a gap causes sparking or a gap short circuit that could damage the tool and workpiece.

4.1.6 Process characteristics

As shown in Fig. 4.9, ECM relies mainly on the ECD phase that occurs by the movement of ions between the cathodic tool and the anodic workpiece.

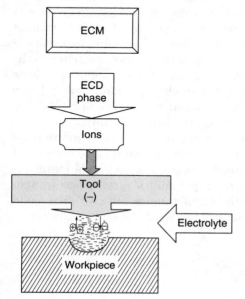

Figure 4.9 ECM process components.

4.1.6.1 Material removal rate.
Faraday's laws describe the rate of material removal. For pure metals, the specific removal rate q_c [mm³/(min · A)] is given by Kaczmarek (1976):

$$q_c = \frac{60\varepsilon}{96{,}500\rho}$$

For a machining current I, the volumetric removal rate Q_v (mm³/min) is given by

$$Q_v = q_c I$$

The linear removal Q_l (mm/min) is a function of the current density J:

$$Q_l = q_c J$$

where

$$J = \frac{(v - \Delta v)\kappa}{y}$$

The specific removal q_c describes how effectively the machining current is utilized for the material removal from the workpiece. The higher the value, the better is the removal rate per unit ampere. Since most metals dissolve at different valences, the exact determination of the theoretical removal rate becomes difficult due to the random phenomena occurring in the interelectrode gap. Confusion may arise when the incorrect valence state is attributed to the dissolution process. In this regard, at a low potential difference, nickel usually dissolves in a divalent state in nitrate and chloride solutions. This mode of dissolution switches to the trivalent state at higher potential differences. Similarly copper has been observed to dissolve in a single valence in chloride solutions and in a divalent state in nitrate solutions. The apparent valence of copper dissolution changes also with the mode of dissolution. In nitrate and sulfate solutions, copper dissolves with a valence of 2. For conditions of transpassive dissolution, this valence has been found to be between 1 and 2 (1.6 at 60 A/cm²). ECM is mostly carried out on hard materials and alloys. For an alloy containing n components of varying percentages the prediction of the removal rate becomes more difficult. The volumetric removal rate Q_v for that particular alloy is given by Kaczmarek (1976).

$$Q_v = I \sum_{i=1}^{i=n} q_c i$$

The linear removal rate Q_l is given by

$$Q_l = J \sum_{i=1}^{i=n} q_c i$$

For better ECM indices, higher accuracy, and a better surface finish, it is essential to choose the proper chemical composition of the electrolyte and the proper current density. A multicomponent electrolyte should be utilized so that for each element of the alloy there will be a corresponding component in the electrolyte solution. In practical ECM the electrolyte is chosen based on the largest component present in the alloy (Kaczmarek, 1976). Low values of current efficiency may indicate a failure to choose the optimum machining conditions that lead to high removal rates and accurate dimensions. Current losses may occur as a result of changes in the electrolyte properties. A poorly chosen electrolyte forms, either a thin layer of elevated gas or an oxide film on the anode surface. The oxide film is difficult to remove and, therefore, increases the polarization resistance. Figure 4.10 arranges some alloys with respect to the theoretical removal rate assuming a current efficiency of 100 percent. Steel 4340 has the highest rate of material removal, while lithium alloy L605 has the lowest one.

Khayry (1989) suggested nickel as a free-cutting material in ECM since it can be machined at 100 percent current efficiency using a low current density of 25 A/cm^2. If the current density is increased (250 A/cm^2), the efficiency is reduced to 85 to 90 percent due to gas evolution.

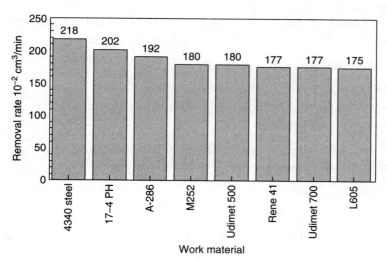

Figure 4.10 Theoretical removal rates for ECM of alloys assuming 100 percent current efficiency and a current density of 1.53 A/mm^2 (El-Hofy, 1995).

When machining titanium, in NaCl electrolyte, a current efficiency of 10 to 20 percent is obtainable. When this solution is replaced by some mixture of florid-based electrolytes, higher efficiencies are obtained at 60 V that break down the tenacious oxide film formed on the machined surface. However, McGeough (1988) reported that this tough, light, and corrosion-resistant layer is what makes titanium so useful in the aircraft engine industry. With low electrolyte flow rates the current efficiency is reduced due to the accumulation of machining products within the gap that impedes further dissolution of the metal. Additionally, the buildup of cathodically generated gases can lead to short circuiting between the tool and workpiece, which terminates the machining process and damages both the tool and workpiece.

Electrolyte concentration is also an important factor that determines the gap resistance and controls the machining current. In some cases the removal rate and hence the machinability can be enhanced by heating the electrolyte because this increases the electrolyte's specific conductivity. In practice, the electrolyte temperature must not exceed 60 to 70°C using temperature regulators.

Electrolyte circulation and replenishment are recommended for raising the electrolyte discharging voltage and keeping the pH number at a moderate level of 4 to 10. The electrolytes used in practice make it possible to employ a working voltage up to 20 V (*Metals Handbook*, 1989). Figure 4.11 shows the relative machinability index of some alloys based on the machining speed obtainable at a constant current density.

4.1.6.2 Accuracy of ECM. A small gap width represents a high degree of process accuracy. As can be seen in Fig. 4.12, the accuracy of machined parts depends on the current density, which is affected by

1. Material equivalent and gap voltage
2. Feed rate and gap phenomena including passivation
3. Electrolyte properties including rate, pH, temperature, concentration, pressure, type, and velocity

For high process accuracy, machining conditions leading to narrow machining gaps are recommended. These include use of

1. A high feed rate
2. High-conductivity electrolytes
3. Passivating electrolytes, such as $NaNO_3$, that have a low throwing power
4. Tool insulation that limits the side-machining action

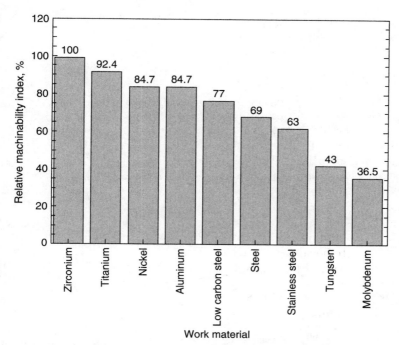

Figure 4.11 Relative machinability for different materials at constant current density (El-Hofy, 1995).

The control of dimensions in ECM can be explained using Fig. 4.13. If the passage of current is only required between RS and BC, a stray current will pass from RS to BA and CD. The current density is less in the directions B to A and C to D. Surface polishing occurs along BC, while the lower current density along BA and CD leads to etching and pitting. Under such conditions, machining takes place in the regions where it is undesired thus leading to a lack of dimensional control. A similar situation occurs in the side gap during electrochemical drilling and die sinking processes.

The extent to which the current strays in this fashion is related to the throwing power of the electrolyte. Fine dimensional control was obtained when using sodium chlorate ($NaClO_3$) electrolyte due to the very low electrolyte throwing power, which is linked to the passivation of the anode in the local current density region. Adding passivating agents such as benzotriazol (Azimidobenzene, $C_6H_5N_3$) and potassium dichromate ($K_2Cr_2O_7$) to NaCl solution reduces the large throwing power of NaCl. Under such conditions, a passive film was found over the machined surface that improved the dimensional control of the machined parts.

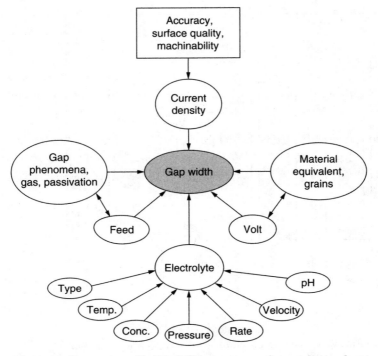

Figure 4.12 Parameters affecting ECM accuracy, surface quality, and productivity.

El-Mehdawy (1977) reported that during machining of steel using NaCl solution, the rate of passive film formation is controlled by the current density and electrolyte flow rate. At a high current density, a faster rate of passive film formation occurs due to the rapid precipitation of the anode products over the machined surface. The increase of electrolyte flow rate reduces the concentration of these products and hence the rate of the film formation.

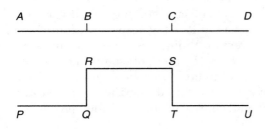

Figure 4.13 Dimensional control in ECM.

The typical dimensional tolerances for ECM are ±0.13 mm for the frontal gap and ±0.25 mm for the side gap. Proper control of the machining parameters leads to a narrow tolerance of ±0.025 mm. It is difficult to machine internal radii smaller than 0.8 mm, and external radii less than 0.5 mm. An overcut of 0.5 mm, taper of 0.001 mm/mm, and corner radii of 2.5 mm are possible and depend on the configuration of the cathodic tool used (*Metals Handbook*, 1989).

4.1.6.3 Surface finish. According to Konig and Lindelauf (1973), considerable variations in surface finish occur due to the workpiece characteristics and machining conditions. Crystallographic irregularities, such as voids, dislocation and grain boundaries, differing crystal structures and orientation, and locally different alloy compositions produce an irregular distribution of current density, thus leaving the microscopic peaks and valleys that form the surface roughness. The mechanism of surface formation can be understood using Fig. 4.14, which shows the effect of machining feed rate on the local gap width for an alloy containing two elements X and Y. Accordingly, due to the difference in their machining rates and their corresponding gap width, the generated maximum peak-to-valley surface roughness R_t decreases at higher feed rates, and thus better surfaces are expected at these higher rates.

The improvement of surface quality, at higher machining currents, shown in Fig. 4.15 was related to the formation of a loose salt layer, which results in a more even distribution of the current density and hence a better surface finish. More fine grained and homogenous structures produce a better surface quality. The roughness obtained with the

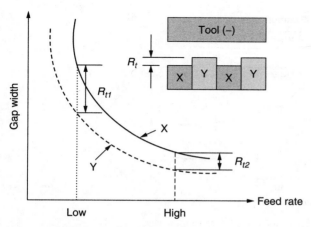

Figure 4.14 Surface roughness generation in ECM.

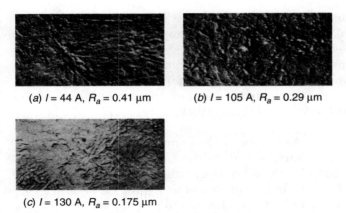

(a) $I = 44$ A, $R_a = 0.41$ μm (b) $I = 105$ A, $R_a = 0.29$ μm

(c) $I = 130$ A, $R_a = 0.175$ μm

Figure 4.15 The appearance of different ECM surfaces at various currents for NC6 heat-treated tool steel (Kaczmarek, 1976).

larger grain size of the annealed carbon steel (Fig. 4.16), was possibly due to the reduced number of grain boundaries present on such a surface.

The surface roughness of a machined part varies from 0.3 to 1.9 μm for the frontal gap area and can be as rough as 5 μm or more for the side gap area. Some metals can be polished in simple salt solutions if the current density is high. Microscopic surface defects such as intergranular attack may be caused by the selective ECM dissolution of certain constituents of the alloy. Such defects are usually associated with low current densities and intermetallic precipitates at grain boundaries. Intergranular attack can usually be avoided by proper selection of electrolyte and machining

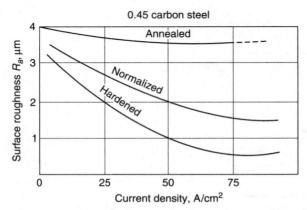

Figure 4.16 Effect of heat treatment on surface roughness for steel (Kaczmarek, 1976).

parameters. In general the following ECM effects are summarized in *Metals Handbook* (1989):

- Large grains cause a rougher finish than fine grains.
- Insoluble inclusions such as graphite in cast iron increase roughness and create machining problems.
- Variations in workpiece composition, as in the case of hardened steel, cause differences in local machining rates.
- Precipitation of intermetallic compounds at grain boundaries leads to serious intergranular attack.

Masuzawa et al. (1991) reported that for nonpassivating electrode systems, the reduction in electrolyte concentration and the increase of its temperature improve the quality of surfaces. For passivating systems, low electrolyte concentration and a rise in its temperature increase the formation of a protective layer that causes deterioration to the surface quality. A further increase in current density breaks up this layer so that there is a decrease in the percentage of the covered surface areas where the reaction of a nonremoving effect occurs and a smoother surface is produced. Etched matt surfaces of roughness 0.9 to 1.4 μm at a current density of 15.5 A/cm^2 and electrolyte flow rates of 28 m/s have been produced on Nimonic 80 using NaCl solution. At a current density of 46.5 A/cm^2 and an electrolyte flow rate of 28.5 m/s, the roughness was 0.2 μm.

The coarse finish with surface roughness of 5.2 μm obtained with cast iron machined in 20% w/w NaCl, current densities ranging from 3.7 to 15.5 A/cm^2, and flow rates of 0.4 to 1.4 m^3/s has been mainly attributed to partial passivation. Marks of partial passivation were also observed by McGeough (1974) in case of 0.78% C quenched and tempered steels at an electrolyte velocity of 16.5 m/s. These marks were broken down as the velocity was increased from 29 to 45 m/s, and hence the surface carried polished striations.

4.1.7 Process control

During ECM many of the selected machining conditions must be kept unchanged as they have a direct impact on the process accuracy and surface finish. The gap voltage affects the produced oversize, which in turn alters the dimensions of the machined workpiece. Surface roughness depends on current density, which is affected by the tool feed rate, gap voltage, work material, gap phenomenon, and electrolyte conditions including type, concentration, temperature, pressure, pH level, and conductivity.

During ECM, machining conditions leading to high process accuracy are associated with smaller surface roughness and greater machining productivity. High current densities are, therefore, recommended.

However, electrolyte heating, and boiling and the increased possibility of sparking should be considered. Proper control of ECM enhances the product quality and process productivity in view of the following reasons:

1. The nonstationary behavior of ECM is due to variations in the interelectrode gap conditions such as gas generation, heating, passivation, and other electrode reactions.
2. For large components, the actual machining time constitutes a high percentage of the total production time. Any improvement in machining performance could yield a significant reduction in product cycle time.
3. The cost of additional control hardware may remain a small fraction of the total machine cost including power supply and electrolyte feeding and filtration units.
4. The increase of unmanned working hours raises the efficiency of ECM machines and enhances the possibility of process integration into CAD/CAM systems.
5. ECM deterioration by sparking may cause irreparable damage to the intricate and often costly workpiece and tools or completely shut down the machine prematurely. This risk rises at high machining speeds which are normally associated with narrow interelectrode gaps. The application of an advanced control system may significantly contribute to the reduction of the risk of such losses.

Since the early 1960s, extensive efforts have been continuously undertaken to understand the complex nature of ECM in two main directions. In this regard, the effects of tool feed rate, gap voltage, and electrolyte conditions on the produced accuracy are investigated. Further research work was considered for process modeling and proper tool design. Datta et al. (1996) and Reddy et al. (1988) were concerned with electrolyte temperature, pressure variations in the interelectrode gap, and the choice of optimum gap voltage that would avoid the occurrence of sparking and the consequent loss of the tool and workpiece.

Begnon and Bedrin (1982) used the eddy current for in-process measurement of the interelectrode gap to adjust some machining parameters. Their work, however, led to the improvement of the accuracy of the produced shapes. A constant gap control system that employs the feed/voltage ratio to achieve constant electrolyte conductivity is introduced in reference British Patent 1037 662. Other ideas of gap control involve the use of a constant feed or controlling the inlet gap pressure. Control of gap pressure avoids sparking and passivation, which are liable to occur when the constant feed technique is used. A further control scheme is described in reference British Patent 937 681, where constant current flow is ensured.

Elsayed (1989) recommended the use of constant current flow to control the machining gap. Setting and maintaining a small stable gap size in pulsed electrochemical machining (PECM) is important for better dimensional control. A high-speed data acquisition system is used to acquire the pulse current signal for developing an on-line monitoring strategy where the signal variance was correlated to the interelectrode gap size by Rajurkar et al. (1995). A control model that considers the dynamic nature of the process and achieves the desired conditions has been introduced by Rajurkar and Schnacker (1988).

Sorkhel and Bhattacharyya (1994) developed a setup with a computer-controlled stepper motor drive control unit for providing variable and automatic feed rates besides autosensing of the ECM gap conditions during machining. Based on the electrical responses received from the machining gap, control of the tool feed rate and electrolyte flow conditions have been considered.

Kozak et al. (1995) introduced an ECM process with numerically controlled movement of the tool electrode. This eliminates the expensive tool shape and raises the machining accuracy. An intelligent knowledge-based system for a ready, on-line knowledge consultancy to select appropriate ECM conditions is described by Amalnik and McGeough (1996) and Khayry (1996).

4.1.8 Applications

ECM has been used in a wide variety of industrial applications ranging from cavity sinking to deburring. The ability to machine high-strength alloys and hardened steel has led to many cost-saving applications where other processes are impractical. Typical applicators for the ECM process are shown in Fig. 4.17.

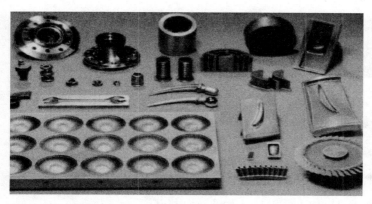

Figure 4.17 ECM components (www.unl.edu/).

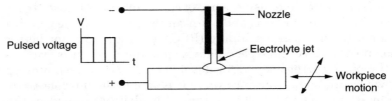

Figure 4.18 Electrolyte jet ECM.

4.1.9 Micro-ECM

ECM is usually characterized as a low-accuracy machining process because of its wider machining gap. In micro-ECM an electrolyte jet is used as a microtool in the arrangement shown in Fig. 4.18. Moving the workpiece and controlling or switching the pulse current produces small indents or cavities in predesigned alignment as shown in Fig. 4.19. For more details regarding micro-ECM, see McGeough (2002).

4.1.10 Advantages and disadvantages of ECM

Advantages

- There is no wear in the tool because there is no contact between the tool and the workpiece.
- Machining is done at low voltages, compared to other processes, with high metal removal rates.
- Very small dimensions up to 0.05 mm can be controlled.
- Complicated profiles can be machined easily in a single operation.

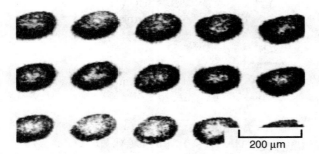

Figure 4.19 Micromachined indents by electrolyte jet ECM (Masuzawa and Tonshof, 1997).

- Because of the low temperature developed, no thermal damage occurs to the workpiece structure.
- Hard conductive materials can be machined.
- The surface finish can be maintained at 0.1 to 1.25 μm R_a.
- Because of its high capital cost, ECM is only suitable for mass production work.
- Labor requirements are low.

Disadvantages

- A huge amount of energy is consumed (about 100 times that required for turning or drilling steel).
- Metal removal rates are slow compared with conventional methods.
- ECM can only be applied to electrically conductive workpiece materials.
- There are difficulties in safely removing and disposing of the explosive hydrogen gas generated during machining.
- The workpiece needs to be cleaned and oiled immediately after machining.
- There are difficulties with handling and containing the electrolyte, which may attack the equipment.
- It is not easy to duplicate the shape of the tool electrode in the workpiece with a high degree of accuracy because of the side machining effect.
- The process can't produce sharp internal or external edges.
- The pumping of high-pressure electrolyte into the narrow machining gap gives rise to large forces acting on the tool and the workpiece.

4.1.11 Environmental impacts

Occupational health dangers are very limited if ECM is applied under appropriate conditions. Electrolyte splashing, contamination of the eyes and skin, and the free expansion of harmful or toxic vapors must be avoided (Fig. 4.20). The crucial impacts on the environment come from the electrolyte and the ECM slurry. During ECM deburring and die sinking, $NaNO_3$ fulfills the requirements of high removal rate and surface quality. However, through the dissolution of metals containing chrome and due to a nitrate reduction at the tool cathode, the medium accumulates toxic chromate and ammonia. These chemical compounds are absorbed into the metal hydroxides in the ECM slurry. Slurries resulting from electrolyte maintenance precipitate solid matters, and solutions of electrolytes have to be disposed off (Tonshof et al., 1996).

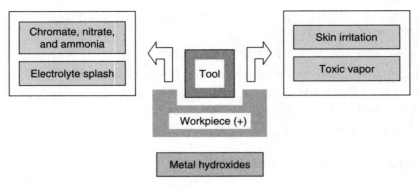

Figure 4.20 Possible impacts of ECM.

4.2 Electrochemical Drilling

Electrochemical drilling (ECDR) produces diameters ranging from 1 to 20 mm, using feed rates from 1 to 5 mm/min. As shown in Fig. 4.21, a tubular electrode is used as the cathodic tool. The electrolyte is pumped from the center of the tool and exits through the side-machining gap, formed between the walls of the tool and the drilled hole. Machining occurs at high current densities in the frontal interelectrode gap between the tool face and the workpiece. Side electrochemical dissolution acts laterally between the sidewalls of the tool and the component. The produced hole diameter is therefore greater than the tool by an overcut C_d, which

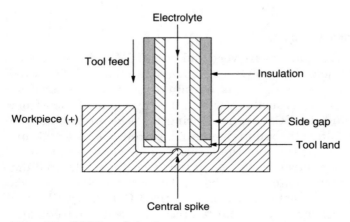

Figure 4.21 ECDR configuration.

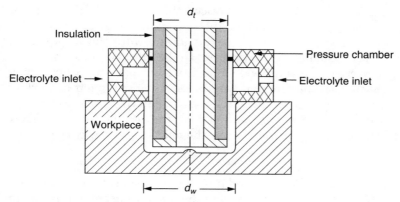

Figure 4.22 Hole drilling using back pressure.

can be calculated, as shown in Fig. 4.22, by

$$C_d = d_w - d_t$$

where d_t is the tool diameter and d_w is the workpiece diameter.

For high machining accuracy and smaller diametral oversize, high feed rates are recommended. Under such conditions, high removal rates and better surface quality are also ensured. The method of electrolyte feeding affects the overcut. In this regard, the reverse electrolyte flow mode under back pressure of 0.6 to 2 MPa, shown in Fig. 4.22, reduces the overcut. This procedure flushes away the gaseous products of electrolysis from the machining gap without reaching the side-machining zone.

The increase in gap pressure raises the electrolyte conductivity, which enhances the dissolution process due to the increase of machining current. The high pressure reduces the size of hydrogen gas bubbles, which according to Rumyantsev and Davydov (1984), raises the electrolyte conductivity. Electrolyte back pressure also eliminates the flow lines in the machined surface. The major disadvantage of such systems, besides the tooling cost, is the increase of hydraulic forces.

Soliman et al. (1986) and Rashed et al. (1986) concluded that the use of proper tool insulation reduces the side-machining effect, which in turn limits the widening of the side gap. Passivating electrolytes such as $NaNO_3$ can produce smaller overcuts, which enhance the process accuracy. The electrolyte flow rate has a pronounced effect on overcut. The use of a rotating tool during ECDR can reduce the roundness error since it ensures homogenous electrolyte flow conditions in the side gap.

In ECDR the passage of current through the interelectrode gap results in the anodic dissolution at a rate that is governed by Faraday's laws of electrolysis. Youssef et al. (1989) noticed that the machining current increases linearly with the tool feed rate. Sparking takes place at a critical feed rate where the speed of tool advance toward the workpiece is greater than the anodic dissolution rate. Under such circumstances, the frontal gap decreases to a critical value at which sparking occurs, causing damage to both the tool and workpiece. The experimental work of Rashed et al. (1976) recommended the use of a gap voltage between 20 to 25 V in order to save energy and reduce the production cost. They described the diametral oversize C_d by the following empirical equation:

$$C_d = 0.225 V^{0.74-0.056a}$$

where V is the gap voltage (V) and a is the tool feed rate (mm min^{-1}).

During ECDR a considerable part of the machining current is lost in the side gap. For this reason the ratio of the experimental removal rate to the theoretical one represents the current efficiency. Current efficiencies lower than 100 percent may be related to gas evolution and the formation of passive oxide film that limits the dissolution process. On the other hand, current efficiencies greater than 100 percent were mainly related to the granular detachment, from the anode, due to the electrolytic grain boundary attack. EC hole drilling is not restricted to circular holes since a tool having any cross section can produce a corresponding shape in the workpiece.

4.3 Shaped Tube Electrolytic Machining

Shaped tube electrolytic machining (STEM) is based on the dissolution process when an electric potential difference is imposed between the anodic workpiece and a cathodic tool. Because of the presence of this electric field the electrolyte, often a sulfuric acid, causes the anode surface to be removed. After the metal ions are dissolved in the solution, they are removed by the electrolyte flow. As shown in Fig. 4.23 and according to

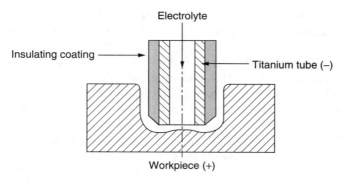

Figure 4.23 STEM schematic.

McGeough (1988), the tool is a conducting cylinder with an insulating coating on the outside and is moved toward the workpiece at a certain feed rate while a voltage is applied across the machining gap. In this way a cylindrically shaped hole is obtained.

STEM is, therefore, a modified variation of the ECM that uses acid electrolytes. Rumyantsev and Davydov (1984) reported that the process is capable of producing small holes with diameters of 0.76 to 1.62 mm and a depth-to-diameter ratio of 180:1 in electrically conductive materials. It is difficult to machine such small holes using normal ECM as the insoluble precipitates produced obstruct the flow path of the electrolyte.

The machining system configuration is similar to that used in ECM. However, it must be acid resistant, be of less rigidity, and have a periodically reverse polarity power supply. The cathodic tool electrode is made of titanium, its outer wall having an insulating coating to permit only frontal machining of the anodic workpiece. The normal operating voltage is 8 to 14 V dc, while the machining current reaches 600 A. The *Metals Handbook* (1989) reports that when a nitric acid electrolyte solution (15% v/v, temperature of about 20°C) is pumped through the gap (at 1 L/min, 10 V, tool feed rate of 2.2 mm/min) to machine a 0.58-mm-diameter hole with 133 mm depth, the resulting diametral overcut is 0.265 mm, and the hole conicity is 0.01/133.

The process also uses a 10% concentration sulfuric acid to prevent the sludge from clogging the tiny cathode and ensure an even flow of electrolyte through the tube. A periodic reversal of polarity, typically at 3 to 9 s prevents the accumulation of the undissolved machining products on the cathode drill surface. The reverse voltage can be taken as 0.1 to 1 times the forward machining voltage. In contrast to the EDM, EBM, and LBM processes, STEM does not leave a heat-affected layer, which is liable to develop microcracks (*www.win.tue.nl/*).

Process parameters

Electrolyte
    Type   Sulfuric, nitric, and hydrochloric acids
    Concentration   10–25% weight in water
    Temperature   38°C (sulfuric acid)
                               21°C (others)
    Pressure   275–500 kPa

Voltage
    Forward   8–14 V
    Reverse   0.1–1 times the forward

Time
- Forward   5–7 s
- Reverse   25–77 ms

Feed rate   0.75–3 mm/min

Process capabilities

Hole size   0.5–6 mm diameter at an aspect ratio of 150
Hole tolerances   0.5-mm diameter ±0.050 mm
                  1.5-mm diameter ±0.075 mm
                  60-mm diameter ±0.100 mm
                  Hole depth ±0.050 mm

Because the process uses acid electrolytes, its use is limited to drilling holes in stainless steel or other corrosion-resistant materials in jet engines and gas turbine parts such as

- Turbine blade cooling holes
- Fuel nozzles
- Any holes where EDM recast is not desirable
- Starting holes for wire EDM
- Drilling holes for corrosion-resistant metals of low conventional machinability
- Drilling oil passages in bearings where EDM causes cracks

Figure 4.24 shows the shape of turbulators that are machined by intermittent drill advance during STEM. The turbulators are normally used for enhancing the heat transfer in turbine engine-cooling holes (*www.win.tue.nl/*).

Advantages

- The depth-to-diameter ratio can be as high as 300.
- A large number of holes (up to 200) can be drilled in the same run.
- Nonparallel holes can be machined.
- Blind holes can be drilled.
- No recast layer or metallurgical defects are produced.
- Shaped and curved holes as well as slots can be produced.

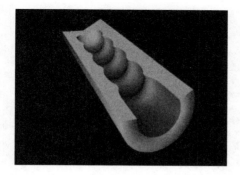

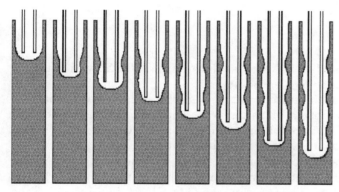

Figure 4.24 Turbulated cooling holes produced by STEM (www.win.tue.nl/).

Limitations

- The process is used for corrosion-resistant metals.
- STEM is slow if single holes are to be drilled.
- A special workplace and environment are required when handling acid.
- Hazardous waste is generated.
- Complex machining and tooling systems are required.

4.4 Electrostream (Capillary) Drilling

Electrostream (ES) (capillary) drilling is a special ECM technique used for producing fine holes that are too deep to produce by EDM and too small to drill by STEM. The cathodic tool used is made from a glass nozzle (0.025–0.50 mm diameter), which is smaller than the

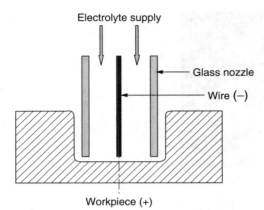

Figure 4.25 ES drilling schematic.

required diameter. The ES process differs from STEM, which uses a coated titanium tube as a cathodic tool.

To conduct the machining current through the acid electrolyte that fills the interelectrode gap, a platinum wire electrode is fitted inside the glass nozzle (Fig. 4.25). Solutions of sulfuric, nitric, or hydrochloric acid with a concentration of 12 to 20 wt % are common electrolytes used. The type electrolyte used depends on the condition of the workpiece. In this regard, hydrochloric solution is used for aluminum and its related alloys, while sulfuric acid solution is recommended for Hastelloy, Inconel, Rene alloys, and carbon and stainless steels.

The electrolyte temperature is normally 40°C for sulfuric acid and 20°C for the rest. Electrolyte pressures between 275 and 400 kPa are recommended. During machining, the electrolyte stream is negatively charged and squirted against the anodic workpiece. The acid temperature, pressure, concentration, and flow rate must be carefully monitored for satisfactory machining. A gap voltage of 70 to 150 V is employed, which is 10 times greater than that of normal ECM.

Banard (1978) drilled small rows of cooling holes (0.127–1.27 mm diameter) in turbine blades, with depth-to-diameter ratios up to 50 to 1, which is higher than those produced by conventional drilling. The process is also used to produce cooling ducts running at an angle of 45° to the surface of the blade and having a diameter less than 0.8 mm as shown in Figs. 4.26 and 4.27. Wire EDM start holes of less than 0.5 mm can also be drilled using ES.

Process capabilities. Feed rates for ES drilling range from 0.75 to 2.5 mm/min. The feed rate depends on the material to be machined irrespective of the number of holes to be drilled simultaneously. Similar to ECM, higher removal rates are associated with larger feed rates and

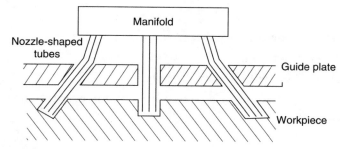

Figure 4.26 Drilling inclined holes.

tool diameters. Additionally, higher removal rates have been reported for titanium alloys than for steel. Normal tolerances are within ±10 percent of the produced hole diameter. The normal hole depth tolerance is ±0.05 mm. These tolerance levels can further be reduced using special control and with pure metals.

Advantages

- High depth-to-diameter ratios are possible.
- Many holes can be drilled simultaneously.
- Blind and intersecting holes can be machined.
- There is an absence of recast and metallurgical defects.
- Powder metallurgy hard materials can be tackled.
- Burr-free holes are produced.

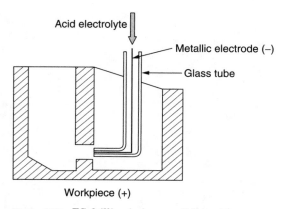

Figure 4.27 ES drilling in inaccessible positions.

Limitations

- Can only be used with corrosion-resistant metals.
- Hazardous waste is generated.
- The process is slow when drilling a single hole.
- The handling of acid requires a special environment and precautions.
- Oblique entry is difficult.

4.5 Electrochemical Jet Drilling

Electrochemical jet drilling (ECJD) is mainly used for fine hole drilling at a diameter-to-depth ratio of 1:12 which is lower than that obtained in ES (capillary) drilling (1:100). As shown in Fig. 4.28, the process does not require the entry of the tool as in the case of ES drilling. The process, therefore, avoids the use of fragile tooling. The jet of a dilute acid electrolyte causes dissolution, and enough room is required for the electrolyte to exit, preferably in the form of spray. A typical voltage in the range of 400 to 800 V is considered optimum.

According to Banard (1978), the lower limit of the hole size is determined by the smallest hole that can be drilled in the cathodic nozzle, the pressure required to pump the electrolyte in the form of a jet, and the amount of overcut. The produced hole diameter depends on the electrolyte throwing power. Generally, holes produced by ECJD are four times the diameter of the electrolyte jet. In ES capillary drilling the hole diameter/capillary diameter ratio is normally less than 2. He reported a ratio of 3:4 using 0.1 M HCl, 0.5-mm jet diameter, 5-mm nozzle distance from workpiece, and an electrolyte pressure of 7×10^5 Nm^{-2}.

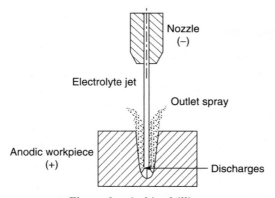

Figure 4.28 Electrochemical jet drilling.

The taper for ECJD is about 5 to 10 included angles, whereas tapers of zero are obtainable in ES capillary drilling.

4.6 Electrochemical Deburring

When machining metal components, it is necessary to cross-drill holes to interconnect bores. Hydraulic valve bodies are a typical example where many drilled passages are used to direct the fluid flow. The intersection of these bores creates burrs, which must be removed (Fig. 4.29) to avoid the possibility of them breaking off and severely damaging the system. Figure 4.30 shows conventionally cut parts that require deburring. Manual removal of burrs is tedious and time-consuming. In the 1970s the thermal energy method (TEM) was introduced to remove burrs in hard-to-reach places. In this method, burrs are hit by 2760°C blast of heat for milliseconds, which burns them away, leaving everything else including threads, dimensions, surface finish, and the physical properties of the part intact. Parts subjected to TEM should be cleaned of oil and metal chips to avoid the formation of carbon smut or the vaporization of chips.

Burrs can be removed using several other methods including vibratory and barrel finishing, tumbling, water blasting, and the application of ultrasound and abrasive slurry. Abrasive flow machining (AFM) provides a reliable and accurate method of deburring for the aerospace and medical industries. AFM can reach inaccessible areas and machine multiple holes, slots, or edges in one operation. It was originally devised in the 1950s for deburring of hydraulic valve spools and bodies and polishing of extrusion dies. The drawbacks of these methods include lack of reliability, low metal removal rates, and contamination of surfaces with grit.

In electrochemical deburring (ECDB), the anodic part to be deburred is placed in a fixture, which positions the cathodic electrode in close proximity to the burrs. The electrolyte is then directed, under pressure, to the gap between the cathodic deburring tool and the burr. On the application of the machining current, the burr dissolves forming a controlled radius. Since the gap between the burr and the electrode

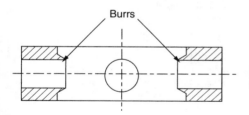

Figure 4.29 Burrs formed at intersections of holes.

Figure 4.30 Different components that require deburring (www.vectron.cc/).

is minimal, burrs are removed at high current densities. ECDB, therefore, changes the dimensions of the part by removing burrs leaving a controlled radius. Figure 4.31 shows a typical EC hole deburring arrangement. ECDB can be applied to gears, spline shafts, milled components, drilled holes, and punched blanks. The process is particularly efficient for hydraulic system components such as spools, and sleeves of fluid distributors.

Mechanism of deburring. Faraday's laws of electrolysis dictate how the metal is removed by ECDB. The deburring speed may be as high as 400 to 500 mm/min. ECDB using a rotating and feeding tool electrode (Fig. 4.32) enhances the deburring process by creating turbulent flow in the interelectrode gap. The spindle rotation is reversed to increase the electrolyte turbulence. Normal cycle times for deburring reported by Brown (1998) are between 30 to 45 s after which the spindle is

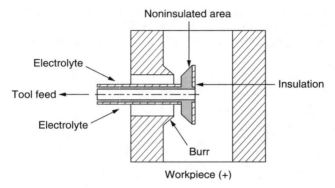

Figure 4.31 Hole deburring.

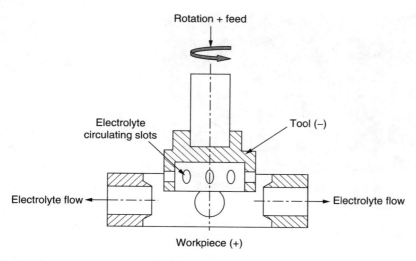

Figure 4.32 Electrochemical deburring using a rotating tool.

retracted and the part is removed. In simple deburring when the tool is placed over the workpiece, a burr height of 0.5 mm can be removed to a radius of 0.05 to 0.2 mm leaving a maximum surface roughness of 2 to 4 µm.

When burrs are removed from intersections of passages in a housing, the electrolyte is directed and maintained under a pressure of 0.3 to 0.5 MPa using a special tool. That tool has as many working areas as practical so that several intersections are deburred at a time. Proper tool insulation guarantees the flow of current in areas nearby the burr. The deburring tool should also have a similar contour of the work part thus leaving a 0.1 to 0.3 mm interelectrode gap. Moreover the tool tip should overlap the machined area by 1.5 to 2 mm in order to produce a proper radius. The choice of the electrolyte plays an important role in the deburring process. Table 4.2 presents different electrolytes and the operating

TABLE 4.2 Machining Conditions for Deburring of Different Materials

Material	Electrolyte	Applied voltage, V	Current density, A/cm^2	Time, s
Carbon steel and low-carbon steel	5–15% NaNO$_3$ 2–2.5 NaNO$_2$	12–24	5–10	5–100
Copper alloys	5–15% NaNO$_3$			
Aluminum alloys	5–20% NaNO$_3$			
Stainless steels	5% NaNO$_3$ +0.5% NaCl			

SOURCE: Rumyantsev and Davydov (1984).

Before   After

Figure 4.33 EC deburring applications (www.extrudehone.com/cation).

conditions for ECDB of some materials. ECDB power units supply a maximum current of 50 A. However, power units having 500 A are used to remove burrs generated by turning and facing operations on large forged parts. Figure 4.33 shows an EC deburring application.

Advantages

- Elimination of costly hand deburring
- Increase of product quality and reliability
- Ensures the removal of burrs at the required accuracy, uniformity, proper radius, and clean edge
- Reduced personnel and labor cost
- Can be automated for higher productivity

References

Amalnik, M. S., and McGeough, J. (1996). "Intelligent Concurrent Manufacturability Evaluation for Design for Electrochemical Machining," *Journal of Material Processing Technology*, 61:130–139.
British Patent 937 681. A system for maintaining a constant current flow between tool and workpiece.
British Patent 1037 662. Control of the ratio of feed rate to the applied voltage in order to achieve a constant predetermined working gap.
Banard, J. (1978). "Fine Hole Drilling Using Electrochemical Machining," *MTDR Conf. Proc.*, pp. 503–510.
Begnon, C., and Bedrin, C. (1982). "Application of Eddy Currents to the In Process Measurement of the Gap in ECM," *Anals of CIRP*, 31(1):115–118.
Brown, J. (1998). *Advanced Machining Technology Handbook*. New York: McGraw-Hill.

Datta, M, Shenoy, R. V., and Romatkiw, L. T. (1996). "Recent Advances in the Study of Electrochemical Micromachining," ASME, *Journal of Engineering for Industry*, 118:29–36.
El-Hofy, H. (1995). "Machinability Indices for Some Non-conventional Machining Processes," *Alexandria Engineering Journal* (AEJ). 34(3):231–245.
El-Hofy, H. (1996). "Surface Generation in Non-conventional Machining," *MDP-6, Conf.*, Cairo, pp. 203–213.
El-Mehdawy, Y. (1977). Accuracy of Electrolytic Sinking, M.S. Thesis, Alexandria University.
El-Sayed, E. M. (1989). Towards Better Control of Electrochemical Machining (ECM) Parameters by Computer Interfacing, M.S. Thesis, King Saud University, Riyadh.
Kaczmarek, J. (1976). *Principles of Machining by Cutting, Abrasion and Erosion*, Stevenage, U.K.: Peter Peregrines Ltd.
Khayry, A. (1989). "Towards a Machinability System for Electrochemical Machining." *CAPE-5 Conf. Proc.*, Edinburgh, pp. 343–350.
Khayry, A. (1996). "A Knowledge-Based System for Electrochemical Machining Procedure," *Journal of Material Processing Technology*, 58:121–130.
Konig, W., and Lindelauf, P. (1973). "Surface Generation in Electrochemical Machining," *Annals of CIRP*, 29(1):97–100.
Kozak, J., Dabrowiski, L., Rusazj, A., and Slawinski, R. (1995). "Computer Simulation on Numerically Controlled Electrochemical Machining (ECM-NC) with a Spherical Tool Electrode," *CAPE-11 Conf.*, pp. 205–210.
Machining Data Handbook. (1997). 3d. ed., Vol. 2. Cincinnati, OH: Machinability Data Center, Institute of Advanced Manufacturing.
Masuzawa, T., and Tonshof, H. K. (1997). "Three Dimensional Micro Machining by Machine Tools," *Annals of CIRP*, 42(2):621–628.
Masuzawa, T., et al. (1991). "Electrochemical Surface Finishing of Tungsten Carbide Alloys," *Annals of CIRP*, 40(1):199–202.
McGeough, J. (1974). *Principles of Electrochemical Machining*. London: Chapman and Hall.
McGeough, J. (1988). *Advanced Methods of Machining*, London. New York: Chapman and Hall, London.
McGeough, J. (2002). *Micromachining of Engineering Materials*. New York: Marcel Dekker, Inc.
Metals Handbook. (1989). Vol. 16, *Machining*, Materials Park, OH: ASM International.
Rajurkar, K. P., and Schnacker, C. L. (1988). "Some Aspects of ECM Performance and Control," *Annals of CIRP*, 37(1):183–186.
Rajurkar, K. P., Wei, B., Kozak, J., and McGeough, J. (1995). "Modeling and Monitoring Inter Electrode Gap in Pulse Electrochemical Machining," *Annals of CIRP*, 44(1):177–180.
Rashed, A. F., Youssef, H., and El-Mehdawy, Y. (1976). "Effect of Gap Voltage and Current in Electrolytic Sinking," *Bulletin of the Faculty of Engineering*, Alexandria University, XV-1:705–736.
Rashed, A. F., Youssef, H., and El-Hofy, H. (1986). "Effect of Some Process Parameters on the Side Machining During Electrolytic Sinking," *PEDAC-3 Conf.*, Alexandria, pp. 733–746.
Reddy, M. S., Jain, V. K., and Lai, G. K. (1988). "Tool Design for ECM: Correction Factor Method," ASME, *Journal of Engineering for Industry*, 110:111–118.
Rumyantsev, E., and Davydov, A. (1984). *Electrochemical Machining of Metals*. Moscow: Mir Publishers.
Soliman, S. M., Youssef, H., and El-Hofy, H. (1986). "A Study of the Side Machining in Electrolytic Sinking," *PEDAC-3 Conf.*, Alexandria, pp. 717–733.
Sorkhel, S. K., and Bhattacharyya, B. (1994). "Parametric Control for Optimal Quality of Workpiece Surface in ECM," *Journal of Material Processing Technology*, 40:271–286.
Tipton, H. (1971). "The Determination of the Shape of Tools for Use in Electrochemical Machining," MTIRA Report No. 40.
Tonshof, K., Egger, K. R., and Kloclco, F. (1996). "Environmental and Safety Aspects of Electro Physical and Electrochemical Processes," *Annals of CIRP*, 45(2):553–567.

Youssef, H., El-Hofy, H., and El-Mehdawy, Y. (1989). "Sparking Phenomena and Hole Oversize by ECM, Effect of Some Process Parameters," *Alexandria Engineering Journal (AEJ)*, 28(4):247–259.
www.extrudehone.com/cation
www.unl.edu/nmrc/ecm1.htm
www.vectron.cc/electro.html
www.win.tue.nl/~scg/drilling/noot.html#part2

Chapter 5

Thermal Processes

5.1 Electrodischarge Machining

5.1.1 Introduction

The history of electrodischarge machining (EDM) dates back to the days of World Wars I and II when B. R. and N. I. Lazarenko invented the relaxation circuit (RC). Using a simple servo controller they maintained the gap width between the tool and the workpiece, reduced arcing, and made EDM more profitable. Since 1940, die sinking by EDM has been refined using pulse generators, planetary and orbital motion techniques, computer numerical control (CNC), and the adaptive control systems.

During the 1960s the extensive research led the progress of EDM when numerous problems related to mathematical modeling were tackled. The evolution of wire EDM in the 1970s was due to the powerful generators, new wire tool electrodes, improved machine intelligence, and better flushing. Recently, the machining speed has gone up by 20 times, which has decreased machining costs by at least 30 percent and improved the surface finish by a factor of 15. EDM has the following advantages:

1. Cavities with thin walls and fine features can be produced.
2. Difficult geometry is possible.
3. The use of EDM is not affected by the hardness of the work material.
4. The process is burr-free.

5.1.2 Mechanism of material removal

In EDM, the removal of material is based upon the electrodischarge erosion (EDE) effect of electric sparks occurring between two electrodes that are separated by a dielectric liquid as shown in Fig. 5.1. Metal removal

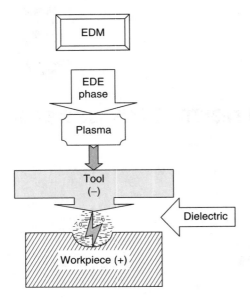

Figure 5.1 EDM components.

takes place as a result of the generation of extremely high temperatures generated by the high-intensity discharges that melt and evaporate the two electrodes. A series of voltage pulses (Fig. 5.2) of magnitude about 20 to 120 V and frequency on the order of 5 kHz is applied between the two electrodes, which are separated by a small gap, typically 0.01 to 0.5 mm. When using RC generators, the voltage pulses, shown in Fig. 5.3, are responsible for material removal.

According to McGeough (1988), the application of voltage pulses, as shown in Fig. 5.4, causes electrical breakdown to the dielectric in a channel of radius 10 μm. The breakdown arises from the acceleration toward the anode of both electrons emitted from the cathode by the

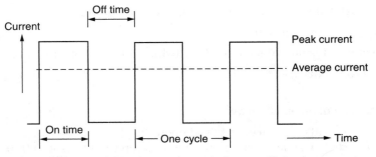

Figure 5.2 Typical EDM pulse current train for controlled pulse generator.

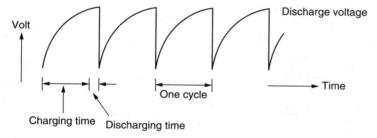

Figure 5.3 Variation of voltage with time using an RC generator.

applied field and the stray electrons present in the gap. These electrons collide with neutral atoms of the dielectric, thereby creating positive ions and further electrons, which in turn are accelerated respectively toward the cathode and anode.

When the electrons and the positive ions reach the anode and cathode, they give up their kinetic energy in the form of heat. Temperatures of about 8000 to 12,000°C and heat fluxes up to 10^{17} W/m^2 are attained. With a very short duration spark of typically between 0.1 to 2000 µs the temperature of the electrodes can be raised locally to more than their

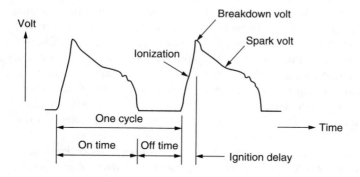

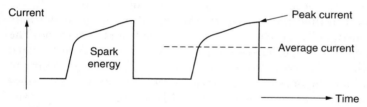

Figure 5.4 Voltage and current waveforms during EDM.

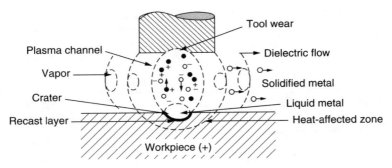

Figure 5.5 EDM spark description.

normal boiling points. Owing to the evaporation of the dielectric, the pressure on the plasma channel rises rapidly to values as high as 200 atmospheres. Such great pressures prevent the evaporation of the superheated metal.

At the end of the pulse, the pressure drops suddenly and the superheated metal evaporates explosively. Metal is thus removed from the electrodes as shown in Fig. 5.5.

Fresh dielectric fluid rushes in, flushing the debris away and quenching the surface of the workpiece. Unexpelled molten metal solidifies to form what is known as the recast layer. The expelled metal solidifies into tiny spheres dispersed in the dielectric liquid along with bits from the electrode. The remaining vapor rises to the surface. Without a sufficient off time, debris would collect making the spark unstable. This situation creates an arc, which damages the electrode and the workpiece. The relation between the amount of material removed from the anode and cathode depends on the respective contribution of the electrons and positive ions to the total current flow. The electron current predominates in the early stages of the discharge. Since the positive ions are roughly 10^4 times more massive than electrons, they are less easily mobilized than the electrons. Consequently the erosion of the anode workpiece should be greater than that of the cathode. At the end of the EDM action, the plasma channel increases in width, and the current density across the interelectrode gap decreases. With the fraction of the current due to the electrons diminishing, the contributions from the positive ions rise, and proportionally more metal is then eroded from the cathode. The high frequency of voltage pulses supplied, together with the forward servo-controlled tool motion, toward the workpiece, enables sparking to be achieved along the entire length of the electrodes. Figure 5.4 shows the voltage and current waveforms during EDM. Figure 5.6 shows the periodic discharges occurring when using an RC generator in EDM.

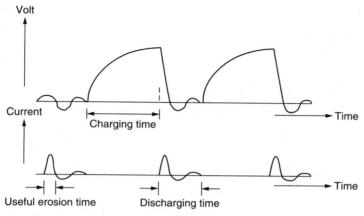

Figure 5.6 Periodic discharges in RC-type generator.

The frequency of discharges or sparks usually varies between 500 and 500,000 sparks per second. With such high sparking frequencies, the combined effects of individual sparks provide a substantial material removal rate. The position of the tool electrode is controlled by the servomechanism, which maintains a constant gap width (200–500 μm) between the electrodes in order to increase the machining efficiency through active discharges.

EDM performance measures such as material removal rate, electrode tool wear, and surface finish, for the same energy, depends on the shape of the current pulses. Based upon the situation in the interelectrode gap, four different electrical pulses are distinguished, namely, open circuit pulses, sparks, arcs, and short circuits. They are usually defined on the basis of time evolution of discharge voltage and/or discharge current. Their effect upon material removal and tool wear differs quite significantly. Open gap voltages that occur when the distance between both electrodes is too large obviously do not contribute to any material removal or electrode tool wear. When sudden contact occurs between the tool and workpiece, micro short circuits occur, which do not contribute to the material removal process. The range of the electrode distance between these two extreme cases forms the practical working gap for actual discharges, i.e., sparks and arcs. In this regard, arcs are believed to occur in the same spot on the electrode surface and may, therefore, severely damage the tool and the workpiece. It is assumed that arcs occur when the plasma channel of the previous pulse is not fully deionized; the current during the following pulse will flow by preference along the same current path. In such a case, no time is required to form new gaseous current paths that initiate a new spark

120   Chapter Five

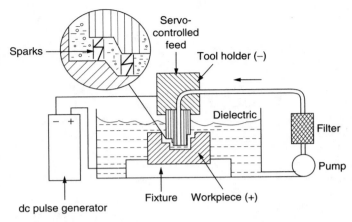

Figure 5.7   EDM schematic.

discharge. It is believed that only "sparks" really contribute to material removal in a desired mode.

5.1.3 The machining system

Figures 5.7 and 5.8 show the main components of the EDM system. These components include the tool feed servo-controlled unit, which maintains a constant machining gap that ensures the occurrence of active discharges between the two electrodes. The power supply is responsible for supplying pulses at a certain voltage, current, on time,

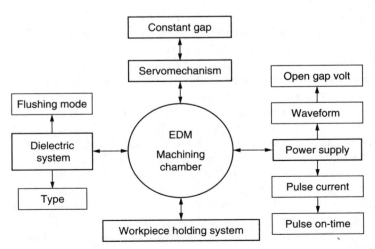

Figure 5.8   EDM system components.

and off time. The dielectric circulation unit flushes the dielectric fluid to the interelectrode gap after being filtered from the machining debris.

5.1.3.1 EDM electrodes

Material. Metals with a high melting point and good electrical conductivity are usually chosen as tool materials for EDM. Graphite is the most common electrode material since it has fair wear characteristics and is easily machinable and small flush holes can be drilled into graphite electrodes. Copper has good EDM wear and better conductivity. It is generally used for better finishes in the range of 0.5 µm R_a. Copper tungsten and silver tungsten are used for making deep slots under poor flushing conditions especially in tungsten carbides. It offers high machining rates as well as low electrode wear. Copper graphite is good for cross-sectional electrodes. It has better electrical conductivity than graphite while the corner wear is higher. Brass ensures stable sparking conditions and is normally used for specialized applications such as drilling of small holes where the high electrode wear is acceptable (*Metals Handbook*, 1989).

Movements. In addition to the servo-controlled feed, the tool electrode may have an additional rotary or orbiting motion. Electrode rotation helps to solve the flushing difficulty encountered when machining small holes with EDM. In addition to the increase in cutting speed, the quality of the hole produced is superior to that obtained using a stationary electrode. Electrode orbiting produces cavities having the shape of the electrode. The size of the electrode and the radius of the orbit (2.54-mm maximum) determine the size of the cavities. Electrode orbiting improves flushing by creating a pumping effect of the dielectric liquid through the gap.

Polarity. Electrode polarity depends on both the workpiece and electrode materials. Table 5.1 shows the possible electrode polarity for different workpiece and tool combinations.

TABLE 5.1 Electrode Polarities for Different Workpiece Materials

Electrode material	Work material				
	Steel	Tungsten carbide	Copper	Aluminum	Ni-base alloys
Graphite	+, −	−	−	+	+, −
Copper	+	+, −	−	+	+
Cu-W	+	+, −	−	+	+
Steel	+, −	+	−	−	−
Brass	−	−		+	−

SOURCE: *Metals Handbook* (1989).

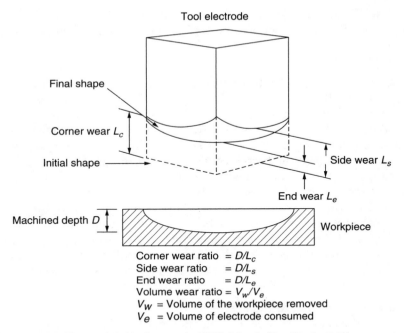

Figure 5.9 Types of electrode wear in EDM (Metals Handbook, 1989).

Electrode wear. The melting point is the most important factor in determining the tool wear. Electrode wear ratios are expressed as end wear, side wear, corner wear, and volume wear as shown in Fig. 5.9. The term *no wear EDM* occurs when the electrode-to-workpiece wear ratio is 1 percent or less. Electrode wear depends on a number of factors associated with the EDM, like voltage, current, electrode material, and polarity. The change in shape of the tool electrode due to the electrode wear causes defects in the workpiece shape. Electrode wear has even more pronounced effects when it comes to micromachining applications. As can be seen from Fig. 5.10 the corner ratio depends on the type of electrode. The low melting point of aluminum is associated with the highest wear ratio. Graphite has shown a low tendency to wear and has the possibility of being molded or machined into complicated electrode shapes. The wear rate of the electrode tool material W_t and the wear ratio R_w, described by Kalpakjian (1997), are

$$W_t = (11 \times 10^3) i T_t^{-2.38}$$

$$R_w = 2.25 \, T_r^{-2.3}$$

where W_t = wear rate of the tool, mm³/min
i = EDM current, A
T_t = melting point of the tool electrode, °C
T_r = ratio of the workpiece to tool electrode melting points

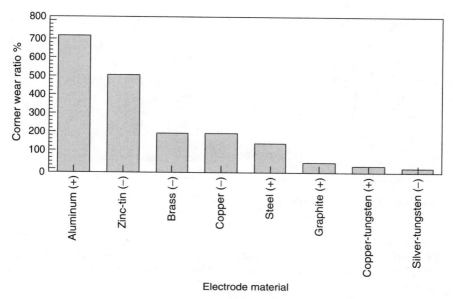

Figure 5.10 Corner wear ratios for different electrode materials (El-Hofy, 1995).

5.1.3.2 Dielectric fluids. The main functions of the dielectric fluid are to

1. Flush the eroded particles from the machining gap
2. Provide insulation between the electrode and the workpiece
3. Cool the section that was heated by the discharging effect

The main requirements of the EDM dielectric fluids are adequate viscosity, high flash point, good oxidation stability, minimum odor, low cost, and good electrical discharge efficiency (*www.unl.edu/nmrc/*). For most EDM operations kerosene is used with certain additives that prevent gas bubbles and de-odoring. Silicon fluids and a mixture of these fluids with petroleum oils have given excellent results. Other dielectric fluids with a varying degree of success include aqueous solutions of ethylene glycol, water in emulsions, and distilled water.

Flushing of the dielectric plays a major role in the maintenance of stable machining and the achievement of close tolerance and high surface quality. Inadequate flushing can result in arcing, decreased electrode life, and increased production time. Four methods of introducing dielectric fluid to the machining gap are considered at *www.unl.edu/nmrc/*.

Normal flow. In the majority of EDM applications, the dielectric fluid is introduced, under pressure, through one or more passages in the tool and is forced to flow through the gap between the tool and the workpiece.

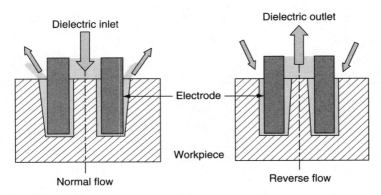

Figure 5.11 Common dielectric flushing modes.

Flushing holes are generally placed in areas where the cuts are deepest. Normal flow is sometimes undesirable because it produces a tapered opening in the workpiece as shown in Fig. 5.11.

Reverse flow. This method is particularly useful in machining deep-cavity dies, where the taper produced using the normal flow mode can be reduced. The gap is submerged in filtered dielectric, and instead of pressure being applied at the source a vacuum is used. With clean fluid flowing between the workpiece and the tool, there is no side sparking and, therefore, no taper is produced as shown in Fig. 5.11.

Jet flushing. In many instances, the desired machining can be achieved by using a spray or jet of fluid directed against the machining gap. Machining time is always longer with jet flushing than with the normal and reverse flow modes.

Immersion flushing. For many shallow cuts or perforations of thin sections, simple immersion of the discharge gap is sufficient. Cooling and machining debris removal can be enhanced during immersion cutting by providing relative motion between the tool and workpiece (Zhixin, 1995). Vibration or cycle interruption comprises periodic reciprocation of the tool relative to the workpiece to effect a pumping action of the dielectric fluid. Synchronized, pulsed flushing is also available on some machines. With this method, flushing occurs only during the nonmachining time as the electrode is retracted slightly to enlarge the gap. Increased electrode life has been reported with this system. Innovative techniques such as ultrasonic vibrations coupled with mechanical pulse EDM, jet flushing with sweeping nozzles, and electrode pulsing are investigated by Masuzawa (1990). For proper flushing conditions, the following was recommended by the *Metals Handbook* (1989):

1. Flushing through the tool is more preferred than side flushing.
2. Many small flushing holes are better than a few large ones.
3. Steady dielectric flow on the entire workpiece-electrode interface is desirable.
4. Dead spots created by pressure flushing, from opposite sides of the workpiece, should be avoided.
5. A vent hole should be provided for any upwardly concave part of the tool-electrode to prevent accumulation of explosive gases.
6. A flush box is useful if there is a hole in the cavity.

5.1.4 Material removal rates

In EDM the metal is removed from both the workpiece and the tool electrode. As can be seen from Fig. 5.12, the material removal rate depends not only on the workpiece material but on the material of the tool electrode and the machining variables such as pulse conditions, electrode polarity, and the machining medium. In this regard a material of low melting point has a high metal removal rate and hence a rougher surface. Typical removal rates range from 0.1 to 400 mm^3/min.

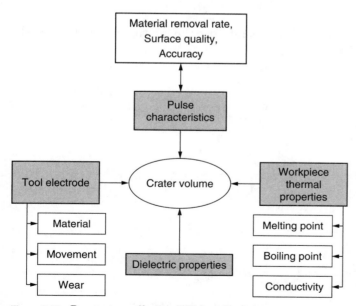

Figure 5.12 Parameters affecting EDM performance.

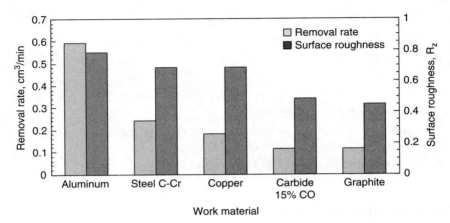

Figure 5.13 EDM removal rates and roughness for different materials (El-Hofy, 1995).

The results of Fig. 5.13 quote machining rates and surface roughness for different materials. Figures 5.14 and 5.15 explain the effect of pulse energy (current) and duration on the crater size and hence the removal rate. The material removal rate, or volumetric removal rate (VRR), in mm³/min, was described by Kalpakjian (1997):

$$\text{VRR} = (4 \times 10^4) i T_w^{-1.23}$$

where i is the EDM current (A) and T_w is the melting point of the workpiece material (°C).

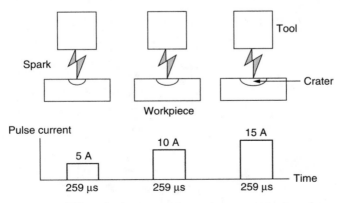

Figure 5.14 Effect of pulse current (energy) on removal rate and surface roughness.

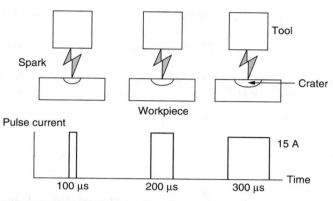

Figure 5.15 Effect of pulse on-time (energy) on removal rate and surface roughness.

5.1.5 Surface integrity

The spark-machined surface consists of a multitude of overlapping crates that are formed by the action of microsecond-duration spark discharges. These craters depend on the physical and the mechanical properties of the material and the composition of the machining medium as well as on the discharge energy and duration as shown in Figs. 5.14 and 5.15. The integral effect of many thousands of discharges per second leads to the formation of the corresponding workpiece profile with a specified accuracy and surface finish. The depth of the resulting craters usually represents the peak to valley (maximum) surface roughness R_t. The maximum depth of the damaged layer can be taken as 2.5 times the average surface roughness R_a. According to Delpreti (1977) and Motoki and Lee (1968) the maximum peak to valley height, R_t, was considered to be 10 times R_a.

The average roughness can be expressed in terms of pulse current i_p (A) and pulse duration t_p (µs) by

$$R_a = 0.0225\, i_p^{0.29}\, t_p^{0.38}$$

A similar observation has been reported by Benhadad et al. (1991) and Dauw and Brown (1990). Stavitskaya and Stravinski (1979) claimed also that the machined surface roughness, which is formed by mutual overlap of craters, is a third of the crater depth. Hence,

$$R_a = \frac{K_h^2 W^{0.33}}{12}$$

where $K_h = 0.4$ to 0.75 µm/µJ$^{0.33}$ and W is the pulse energy (µJ).

Surface roughness increases linearly with an increase in the material removal rate. Jeswani (1978) reported that graphite electrodes produce rougher surfaces than metal ones. Crookall and Khor (1974) related the crater volume/metal removal per discharge to surface roughness H_{rms} while Jeswani (1977) expressed the roughness in terms of the frequency of pulses f_p and power P_r by

$$H_{rms} = 267 \left(\frac{P_r}{f_p}\right)^{0.258}$$

Jeswani added also that as the pulse energy is decreased, the surface finish improves and consequently, the depth at which all formed craters disappear from the machined surface (free polishing depth) is reduced. This depth was found to lie between 3 to 6 times H_{rms} and is important when polishing dies and molds and when the residual stresses are to be removed from the machined surfaces. Masuzawa and Saki (1978) adopted a quick electrochemical finishing technique, using a mate electrode. Accordingly, a reduction of surface roughness from 22 µm R_{max} to 8 µm has been reported together with the removal of the heat-affected layer. In contrast, the matte appearance of the machined surfaces has been found satisfactory in some applications of electrodischarge texturing (EDT).

Kuneida and Furuoya (1991) claimed that the introduction of oxygen gas into the discharge gap provides extra power by the reaction of oxygen, which in turn increased the melting of the workpiece and created greater expulsive forces that increased the metal removal rate and surface roughness. When EDM is used for cusp removal, the silicon powder has been suspended in the working fluid, during the stage of finish EDM. Consequently, Mohri and Saito (1991) reported a change of surface roughness from 45 µm to 10 µm R_{max}. The choice of the correct dielectric flow, in the gap, has a significant effect in reducing the surface roughness by 50 percent, increasing the machining rate, and lowering the thermal effects in the workpiece surface.

The *Machining Data Handbook* (1997) recommended dielectrics having low viscosity for EDM of smooth surfaces. Benhadad et al. (1991) indicated that for Al-Li alloys, the tensile strength of the machined parts are reduced by increased surface roughness. This reduction was enhanced by increased pulse current. The heat-affected layer reached 200 µm compared to 80 µm for steel due to deference in their thermal conductivity. Normal tolerances are about ±25 µm with ±5 µm obtained by proper choice of process variables.

5.1.6 Heat-affected zone

With the temperature of the discharges reaching 8000 to 12,000°C, metallurgical changes occur in the surface layer of the workpiece. Additionally a thin recast layer of 1 μm at 5- μJ powers to 25 μm at high powers is formed. Delpretti (1977) and Levy and Maggi (1990) claimed that the heat-affected zone adjacent to the resolidified layer reaches 25 μm. Some annealing of the workpiece can be expected in a zone just below the machined surface. In addition, not all the workpiece material melted by the discharge is expelled into the dielectric. The remaining melted material is quickly chilled, primarily by heat conduction into the bulk of the workpiece, resulting in an exceedingly hard surface. The depth of the annealed layer is proportional to the amount of power used in the machining operation. It ranges from 50 μm for finish cutting to approximately 200 μm for high metal removal rates. The amount of annealing is usually about two points of hardness below the parent metal for finish cutting. In the roughing cuts, the annealing effect is approximately five points of hardness below the parent metal (Fig. 5.16).

Choosing electrodes that produce more stable machining can reduce the annealing effect. A finish cut removes the annealed material left by the previous high-speed roughing. The altered surface layer, which is produced during EDM, significantly lowers the fatigue strength of alloys. The altered layer consists of a recast layer with or without microcracks, some of which may extend into the base metal, plus metallurgical alterations such as rehardened and tempered layers, heat-affected zones, and intergranular precipitates. Generally, during EDM roughing, the layer showing microstructural changes, including a melted and resolidified layer, is less than 0.127 mm deep, while during EDM finishing, it is less than 0.075 mm. Posttreatment to restore the fatigue strength is recommended to follow EDM of critical or highly stressed surfaces. There are several effective processes that accomplish restoration or even enhancement of the fatigue properties. These methods include the removal of the altered layers by low-stress grinding, chemical machining, and the addition of a metallurgical-type coating, reheat treatment, and the application of shot peening.

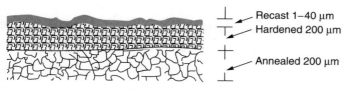

Figure 5.16 EDM heat-affected zones.

5.1.7 Applications

EDM has become an indispensable process in the modern manufacturing industry. It produces complex shapes to a high degree of accuracy in difficult-to-machine materials such as heat-resistant alloys, superalloys, and carbides. The incorporation of EDM within a computer-integrated manufacturing (CIM) system reduced the length of time that the unit operation, without stops for maintenance, is required. Micromachining of holes, slots, and dies; procedures for surface deposition; modification; texturing; milling; and mechanical pulsing are typical applications.

5.1.7.1 Drilling. EDM drilling uses a tubular tool electrode where the dielectric is flushed down the interior hole of the tube in order to remove machining debris. When solid rods are used; the dielectric is fed to the machining zone by either suction or injection through pre-drilled holes. Irregular, tapered, curved, as well as inclined holes can be produced by EDM. McGeough (1988) reported typical feed rates of 0.1 mm/min when drilling 0.1 to 0.5 mm diameters, leaving an overcut of 0.01 to 0.05 mm. Creating cooling channels in turbine blades made of hard alloys is a typical application of EDM drilling. The use of a computerized NC system enabled large numbers of holes to be accurately located.

5.1.7.2 Sawing. ED sawing, shown in Fig. 5.17, is an EDM variation that employs either a special steel band or disc. The process cuts any electrically conductive material at a rate that is twice that of the conventional abrasive sawing method. The cut produced has a smaller kerf besides being free from burrs. Cutting of billets and bars is a typical application.

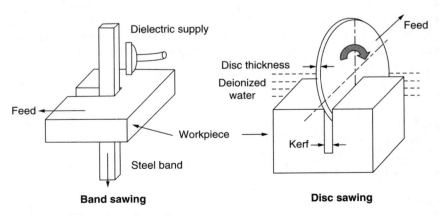

Figure 5.17 ED sawing schematic.

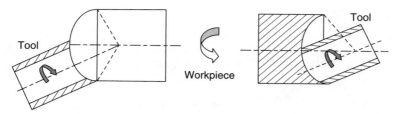

Figure 5.18 Rotary EDM.

A fine finish of 6.3 to 10 μm, leaving a recast layer of 0.025 to 0.130 mm deep, has been reported by the *Machining Data Handbook* (1997).

5.1.7.3 Machining of spheres. Shichun and coworkers (1995) used simple tubular electrodes in the EDM machining of convex and concave spheres, to a dimensional accuracy of ±1 μm and a surface roughness of less than 0.1 μm. Rotary EDM is used for machining of spherical shapes in conducting ceramics using the tool and workpiece arrangement shown in Fig. 5.18.

5.1.7.4 Machining of dies and molds. EDM milling uses standard cylindrical electrodes. Complex cavities are machined by successive NC sweeps of the electrode down to the desired depth. The simple-shaped electrode (Fig. 5.19) is rotated at high speeds and follows specified paths in the workpiece like the conventional end mills. This technique is very useful and makes EDM very versatile like the mechanical milling process. The process solves the problem of manufacturing accurate and complex-shaped electrodes for die sinking of three-dimensional cavities shown in Fig. 5.20. EDM milling enhances dielectric flushing due to the high-speed electrode rotation. The electrode wear can be optimized because of the rotational and contouring motions of the electrode. The main limitation in the EDM milling is that complex shapes with sharp

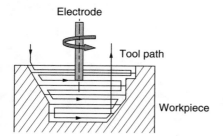

Figure 5.19 EDM die milling.

132 Chapter Five

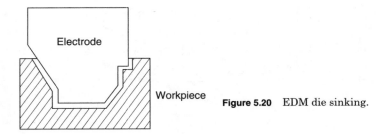

Figure 5.20 EDM die sinking.

corners cannot be machined because of the rotating tool electrode. Figure 5.21 shows the flowchart of the die sinking process. These numerous and time-consuming steps are greatly reduced using EDM milling as shown in Fig. 5.22. EDM milling also replaces the conventional die making that requires the use of a variety of machines such as milling, wire cutting, and EDM die sinking machines.

5.1.7.5 Wire EDM. Wire EDM is a special form of EDM which uses a continuously moving conductive wire electrode. Material removal occurs as a result of spark erosion as the wire electrode is fed, from a fresh wire spool, through the workpiece. In most cases the horizontal movement of the worktable, controlled by CNC, determines the path of the cut, as illustrated in Fig. 5.23. A remarkable application includes the machining

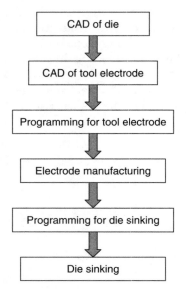

Figure 5.21 EDM die sinking flowchart.

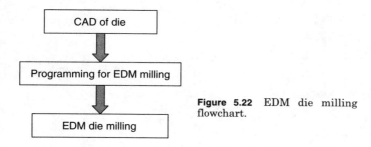

Figure 5.22 EDM die milling flowchart.

of superhard materials such as polycrystalline diamond (PCD) and cubic boron nitride (CBN) blanks, and other matrix composites. Despite the wide use of carbon fiber composites, in the aerospace, nuclear, automobile, and chemical industries, their conventional machining is difficult, owing to delaminating, splintering, burr formation, and short tool life. Kozak et al. (1995) used wire EDM for accurately shaping these materials, without distortion or burrs. The process has been recently used for machining insulating ceramics by Tani et al. (2004).

5.1.7.6 EDM of insulating ceramics. Insulating ceramics are ED-machined by the apparatus shown in Fig. 5.24. A sheet metal mesh is placed over the ceramic material. Spark discharges occur between the negative tool electrode and the metal mesh. These sparks are transmitted through the latter to its interface with the ceramic surface, which is then eroded. A china pot produced by this method is shown in Fig. 5.25.

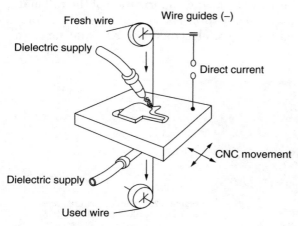

Figure 5.23 Wire EDM schematic.

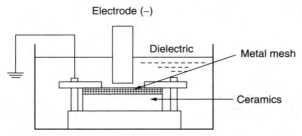

Figure 5.24 EDM of insulating ceramics (Fukuzawa et al., 1995).

5.1.7.7 Texturing. Texturing is applied to the steel sheets during the final stages of cold rolling. Shot blasting (SB) is an inexpensive, conventional texturing method in which hard steel shot is fired at high velocities, roughening the surface on impact. Limitations of SB include its lack of control and consistency of texturing, and the need for protection of other parts of the equipment holding the roll. As the steel rolls become harder, an increasingly limited range of roughness and peak counts have emerged as a further drawback for SB as indicated by Pawleski et al. (1994) and Ahmed and Knight (1988). EDT, is a variation of EDM where the material removal and surface and subsurface modifications are low. EDT proved to be the most popular since the hardness of the roll is not a limitation. Using the new method, texturing is achieved by passing high-intensity electrical sparks of short duration across the gap between the roll (workpiece) and a tool electrode, through which a liquid dielectric (e.g., paraffin) is flushed (Aspinwal et al., 1991). Each spark creates a small crater by the discharge of its energy in a local melting and vaporization of the roll material. By selecting the appropriate process variables such as pulse current, duration and pause times, electrode polarity and material,

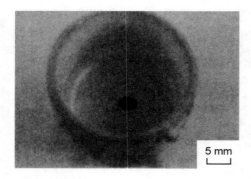

Figure 5.25 China pot machined by EDM. (Fukuzawa et al., 1995).

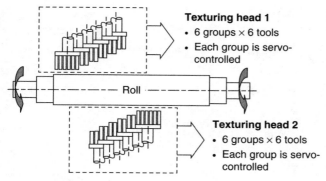

Figure 5.26 EDT machining system.

dielectric type, and the roll rotational speed, a surface texture with a high degree of accuracy and consistency can be produced. The EDT apparatus is shown in Fig. 5.26 where the rotating steel roll is subjected to texturing heads consisting of an array of finger-shaped electrodes. Each finger is electrically isolated from its neighbor and connected to one or more servo-controlled units that can move its position toward, or from, the work roll. McGeough and Rasmussen (1992) used a multichannel power supply to provide controlled sparks usually with a current amplitude of 2 to 10 A and a duration of 10 to 100 µs. An advantage of EDT over SB has been reported by Aspinwal et al. (1989) where the control of the peak count is independent of the surface roughness.

A segmental-type electrode was adopted by El-Menshawy and Ahmed (1985) using a matrix of graphite or copper fingers that are shaped to accommodate the curvature of the work roll. During EDT the workpiece material (steel roll) is rigidly mounted in order to avoid any possible vibrations. The spindle rotates at a rate of 10 to 40 rpm. The rotation of the roll is synchronized with the axial traverse motion of the machining head. Amalnik and coworkers (1997) concluded that the increase in the number of tools raises the machine capital cost, reduces the time required for texturing, and lowers the texturing cost. The increase of pulse current used in the spark gap allows more energy to be dissipated in the plasma channel. Under such conditions, larger craters are formed and the textured surface becomes rougher and hence the number of peaks per unit area also decreases.

5.1.7.8 Micro-EDM. McGeough and De Silva (1996) claimed that EDM can produce holes, normally in the range of 0.1 to 1.0 mm at a depth-to-diameter ratio of 10:1, and at a low machining speed of 1.0 mm/min. There is an increasing demand for greater precision and faster machining. To that end, Chen and McGeough (1994) reported a rapid EDM

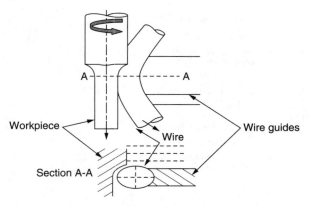

Figure 5.27 Principles of WEDG.

technique for drilling accurate holes, 0.5 to 1.0 mm in diameter and 100 mm deep at a rate of 20 mm/min. Fujino et al. (1995) drilled microholes of diameter 300 µm through tungsten carbide alloy, with a 0.2-mm copper tungsten electrode. Using low-viscosity dielectrics Allen and Huang (1995) described multielectrode drilling of holes with diameters as small as 5 µm. Masuzawa and coworkers (1994) developed a combined electroforming and wire EDM method for micronozzle fabrication. In a further work by Allen and Lecheheb (1996), ink-jet orifices, 75 µm in diameter, were machined in thin stainless steel foils using small energy levels. The principle of wire electrodischarge grinding (WEDG) is illustrated in Fig. 5.27. The material removal mechanism is the same as in EDM. The tool electrode is a metal wire, which slides slowly at 5 to 10 mm/min along a groove in a wire guide. This wire movement makes the process free from the tool wear problems of EDM. By feeding the rotating workpiece in the downward direction, a micropin (Fig. 5.28) has

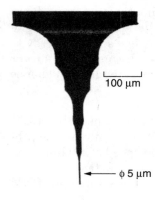

Figure 5.28 Micropin produced by WEDG (Masuzawa and Tonshof, 1997).

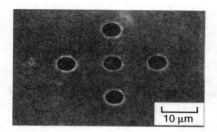

Figure 5.29 Microholes produced by EDM (Masuzawa and Tonshof, 1997).

been produced by Masuzawa and Tonshof (1997). A novel method for micropin fabrication, using machined holes as tools, has been developed by Yamazaki et al. (2004). Micro-EDM machines with WEDG facilities are already in the market, and small hole diameters of 5 μm are available (Fig. 5.29). Using shaped electrodes, holes and cavities with various shapes can also be produced. Figure 5.30 shows a cavity with sharp corners, machined using a 50-μm square section electrode.

5.1.8 Process control

EDM monitoring and control have evolved from fundamental research into modeling single discharges, through stochastic models (dynamic data-dependent systems) to more application-oriented on-line systems such as adaptive control, expert systems, fuzzy logic, and artificial neural networks. Avoidance of abnormal electrical discharges in EDM remains a major difficulty in advances of the process, despite the arrival of the radio-frequency methods devised by El-Menshawy and coworkers (1985). Early EDM adaptive control systems involved basically checking for a "no-load" voltage condition from the discharge gap voltage waveforms. If an electrical discharge without no-load voltage is present for a defined time, the off time for the pulse is increased to avoid these unwanted, abnormal discharges; only when an improvement in machining is achieved is the off time again reduced. Even with modern advances in feedback control, a consistent

Figure 5.30 A microcavity with sharp corners by micro-EDM (Masuzawa and Tonshof, 1997).

quality of EDM has remained difficult to master, at least until the fuzzy adaptive control concepts had been introduced by Lauwers and Kruth (1995). The latter is basically an expert system, performed by a skilled operator.

In brief, the quality of machining is determined from fuzzy rules obtained from experimental knowledge for recognizing instability in machining. Information provided by a signal processor is then used to decide on the appropriate electromachining conditions. Dauw and Van Coppenolle (1995) describe current industrial equipment that incorporates fuzzy controllers. Artificial neural networks are a further new tool in EDM control, especially when combined with fuzzy logic. The neural nets are already used in EDM for monitoring and off-line process planning by Lauwers and Kruth (1995). In a simple application, pulse duration and current are used as inputs, and normal and abnormal (undesired) pulses are classified in the output stage. In the neural net, machining depth, roughness, and accuracy are correlated with machining conditions at the output stage. Industrial versions of neural networks are now available.

5.1.9 EDM automation

The introduction of EDM into advanced manufacturing systems with high levels of computer integration and automated material handling has progressed at a much slower rate than the integration of conventional machining processes such as milling and turning. Recent developments in communications control and CAD/CAM systems for EDM has created the potential for bringing EDM into the mainstream of automated machining operations in modern manufacturing systems. EDM is suited for CIM for many reasons mentioned by Dauw and Van Coppenolle (1995). The level of automation currently exists in EDM units. The relatively low production rate for EDM compared to traditional machining processes is related to the availability of advanced software in the areas of CAD, computer-aided process planning (CAPP), and machining. These three techniques can be dealt with independently, but have to be integrated for complete success of CIM to be guaranteed. There is a growing shortage of highly skilled EDM operators, for whom a fairly good understanding of the physical process is a prerequisite. CIM integration can free EDM from its island of automation, achieving more efficient and cost-effective production. Tool design is performed by the classical approach using drawing rulers and tapes or through CAD systems.

The manufacture of the tool electrode is then performed using several production processes including milling and turning. Finally, the tool arrives at the EDM machine in its chuck holder, perhaps with the assistance of a flexible tool and workpiece transportation system, which in turn may have received its information from a computer-assisted process planner. In a CIM concept, EDM machining centers are integrated on the shop floor with other machining centers. The EDM production machines are integrated into a CAD/CAM system that already exists on the customer's shop floor. The EDM, NC milling, turning, and drilling machines communicate with the central CAD/CAM system by means of a multiuser interface on a local area network (LAN) (Dauw and Van Coppenolle, 1995).

5.1.10 Environmental impact

The environmental impact of machining processes is becoming increasingly significant in view of new legislation in the 1990s. EDM has several hazard potentials, which are described in Fig. 5.31. The high temperature in the working gap results in the generation of hazardous smoke, vapor, and aerosols. Decomposition products and heavy metals may accumulate in the dielectric and the erosion slurry. Hydrocarbon dielectrics have a negative effect on the skin. The sharp-edged metallic particles may damage skin. There is, furthermore, a significant fire hazard. Explosions may occur under unfavorable conditions.

The reduction of electric energy consumption obviously has economic as well as environmental benefits. To that end, research is continuously being done to improve process efficiency by utilizing sophisticated control systems and new types of power generators. EDM die sinking uses an oil-based dielectric, giving an added problem of hydrocarbonaceous fumes, which may be harmful to human health. Research is being

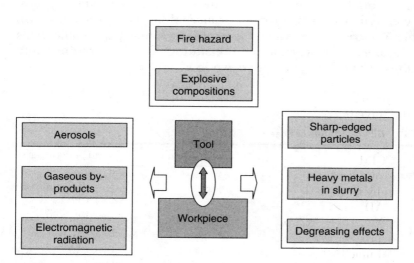

Figure 5.31 EDM hazard potential.

carried out to improve extraction facilities and to reduce emissions by using an alternative dielectric. Unlike die sinking, wire cutting utilizes deionized water as the dielectric. Although the fumes are not as toxic, the long-term effects on the environment due to ozone and chlorine emissions have to be borne in mind. Furthermore, wastewater can prove increasingly difficult to dispose of as environmental laws are further tightened. Environmentally friendly and high-capacity dielectric regeneration for wire EDM has been devised by Levy (1993). The ways to EDM "clean technology" is reviewed by Tonshof et al. (1996).

5.2 Laser Beam Machining

5.2.1 Introduction

Modern machining methods are established to fabricate difficult-to-machine materials such as high-strength thermal-resistant alloys; various kinds of carbides, fiber-reinforced composite materials, Stellites, and ceramics. Conventional machining of such materials produces high cutting forces that, in some particular cases, may not be sustained by the workpiece. Laser beam machining (LBM) offers a good solution that is indeed more associated with material properties such as thermal conductivity and specific heat as well as melting and boiling temperatures.

Laser is the abbreviation of light amplification by stimulated emission of radiation. A highly collimated, monochromatic, and coherent light beam is generated and focused to a small spot. High power densities (10^6 W/mm^2) are then obtained. A large variety of lasers are available in the market including solid-state, ion, and molecular types in either continuous wave (CW) or pulsed mode (PM) of operation as shown in Table 5.2. Lasers are widely used in many industrial applications including plating, heat treatment, cladding, alloying, welding, and machining. The LBM system is shown in Fig. 5.32.

TABLE 5.2 Different Types of Lasers

Laser type		Wavelength, nm	Typical performance
Solid	Ruby	694	Pulsed, 5 W
	Nd-YAG	1064	Pulsed, CW, 1–800 W
	Nd-glass	1064	Pulsed, CW, 2 mW
Semiconductor	GaAs	800–900	Pulsed, CW, 2–10 mW
Molecular	CO_2	10.6 µm	Pulsed, CW, <15 kW
Ion	Ar^+	330–530	Pulsed, CW, 1 W–5 kW
	Excimer	200–500	Pulsed
Neutral gas	He-Ne	633	CW, 20 mW

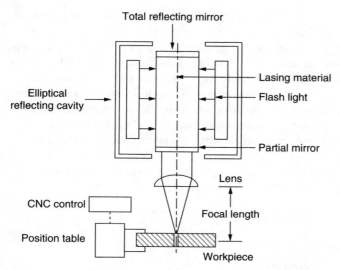

Figure 5.32 LBM schematic.

5.2.2 Material removal mechanism

As presented in Fig. 5.33, the unreflected light is absorbed, thus heating the surface of the specimen. On sufficient heat the workpiece starts to melt and evaporates. The physics of laser machining is very complex

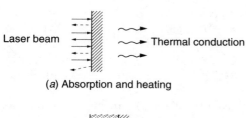

(*a*) Absorption and heating

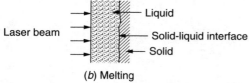

(*b*) Melting

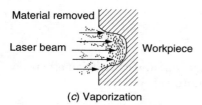

(*c*) Vaporization

Figure 5.33 Physical processes occurring during LBM.

due mainly to scattering and reflection losses at the machined surface. Additionally, heat diffusion into the bulk material causes phase change, melting, and/or vaporization. Depending on the power density and time of beam interaction, the mechanism progresses from one of heat absorption and conduction to one of melting and then vaporization. High-intensity laser beams are not recommended since they form a plasma plume at or near the surface of the material with a consequent reduction in the process efficiency due to absorption and scattering losses.

Machining by laser occurs when the power density of the beam is greater than what is lost by conduction, convection, and radiation, and moreover, the radiation must penetrate and be absorbed into the material. The power density of the laser beam, P_d, is given by

$$P_d = \frac{4L_p}{\pi F_l^2 \alpha^2 \Delta T}$$

The size of the spot diameter d_s is

$$d_s = F_l \alpha$$

The machining rate ϕ (mm/min) can be described as follows:

$$\phi = \frac{C_l L_p}{E_v A_b h}$$

$$A_b = \frac{\pi}{4}(F_l \alpha)^2$$

$$\phi = \frac{4 C_l L_p}{\pi E_v (F_l \alpha)^2 h}$$

The volumetric removal rate (VRR) (mm^3/min) can be calculated as follows:

$$\text{VRR} = \frac{C_l L_p}{E_v h}$$

where P_d = power density, W/cm^2
L_p = laser power, W
F_l = focal length of lens, cm
ΔT = pulse duration of laser, s
α = beam divergence, rad
C_l = constant depending on the material and conversion efficiency
E_v = vaporization energy of the material, W/mm^3

A_b = area of laser beam at focal point, mm^2
h = thickness of material, mm
d_s = spot size diameter, mm

In LBM the workpiece material is removed through several effects including reflection, absorption, and conduction of light that is followed by melting and evaporation. The behavior of the work material with respect to these effects determines the material removal rate. Reflectivity depends on the wavelength, the properties of material, the surface finish, its level of oxidation and temperature.

At a given wavelength, the higher the reflectivity of the material, the lower is the removal rate, and vice versa. In this regard most metals reflect about 90 percent of incident radiation at low power densities. The level of reflectivity can substantially be reduced, and the material machinability is enhanced by modifying the surface conditions. The unreflected light is then absorbed, which in metals raises the electrons to a higher energy state. The energy absorbed by electrons is rapidly passed to the lattice. On sufficient heat the workpiece starts to melt and evaporates. The minimum laser energies required to vaporize plastics are lower than that for metals. The thermal properties of these materials indicate that the machinability is enhanced for materials of low thermal conductivity, diffusivity, and melting point. Most nonmetals have low thermal conductivity and absorb CO_2 laser of 10.6-µm wavelength. Therefore, plastic materials can be readily melted by a low-power CO_2 laser.

Since metals have higher reflectivity and thermal conductivity, nonmetals machine at higher speeds than do metallic ones (Fig. 5.34). Moreover to achieve the same machining speed, greater power density lasers should be

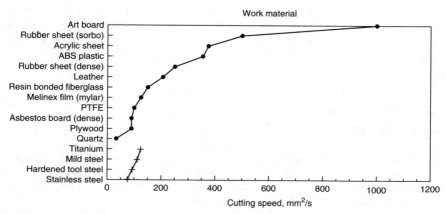

Figure 5.34 Cutting speeds in CO_2 LBM (El-Hofy, 1995).

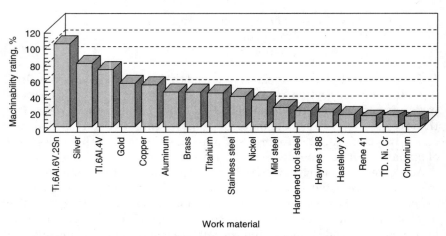

Figure 5.35 Machinability of some materials in LBM (El-Hofy, 1995).

used for metallic materials. The smaller depth of altered layer associated with high machining rates is an advantage of LBM.

In gas-assisted laser the heat of the torch causes the initial melting and oxidation of metal. Machining is achieved by the release of the heat from the oxidation process. The flow of gas stream also contributes by removing the oxide from the machining area. Using such a method, titanium of 0.5 mm thickness has been cut with a CO_2 laser of 135 W at 15 m/min, thus leaving a heat-affected layer of 0.375 mm. A plate of 10-mm thickness of the same material has been also cut at 2.4 m/min, as has copper having 0.025-mm thickness with chlorine as the reactive gas. It should be borne in mind that the smaller the diameter of the gas nozzle and the narrower its distance from the workpiece surface, the better is the quality of the cut. Gas pressure also plays a significant part in the determination of the quality and the rate of machining. Figure 5.35 shows the relative machinability index for a wide range of materials. An index of 100 is taken for Ti.6Al.6V.2Sn, which cuts at the highest speed as shown in Fig. 5.36.

5.2.3 Applications

5.2.3.1 Drilling. Composite materials and some exotic alloys have been widely used in the fabrication of both structural and nonstructural members of the airframes and engines of aircrafts. Superalloys are used frequently for the fabrication of aero-engine gas-path components such as blades, guide vans, after-burners, and castings where a temperature of 2000°C can be reached.

Advanced ceramics and titanium alloys are being developed and used. Such materials need advanced technologies in their fabrication and

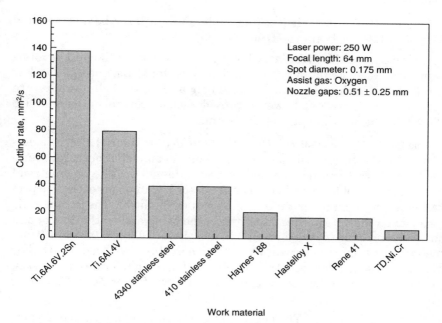

Figure 5.36 Machining speeds of aerospace alloys in gas-assisted LBM (El-Hofy, 1995).

processing. Conventional drilling and punching are found to be insufficient because of the material toughness. Other nonconventional processes such as electrochemical machining (ECM), electrochemical drilling (ECD), EDM, ultrasonic machining (USM), water jet machining (WJM), and abrasive water-jet machining (AWJM) suffer from the high cost of tools and long lead times. The capability of LBM to produce fine holes at high machining speeds compared to other machining processes is reviewed by Snoeys et al. (1986). Since the 1970s, laser drilling has proven to be an efficient drilling tool in terms of cost, quality, and reliability for large-scale production applications. Laser drilling is performed by the following three methods:

A. Direct (percussion) drilling. Produces the required hole size by the application of one or more focused laser pulses. The hole size is determined by the input power and the focusing adjustments. While most holes are drilled with a single laser pulse of high energy, it is more efficient to drill deep holes with multiple pulses of low energy. Under such circumstances Yeo et al. (1994) and Tam et al. (1990) produced better-defined holes.

B. Drill and ream. Consists of drilling a pilot hole and then changing the laser parameters, by increasing the spot size to open out the hole to the required diameter. An alternative way, introduced by Allen and

Huang (1997), is to drill a pilot hole using a laser beam and then apply EDM to obtain the true size.

C. Trepanning. Used for large holes by either moving the part or rotating the laser beam system of delivery or lens. Because of the low energy of each pulse, the recast layer and out-of-cylindericity are reduced. The process, however, provides easier control of the hole size by rotating the diameter of the beam.

Hole quality characteristics. *Diameter and depth.* Using LBM it is economically effective to drill holes up to 1.5 mm in diameter. For large diameters, the trepanning method is recommended. The geometrical characteristics of laser-drilled holes are described in terms of their aspect ratio (depth-to-diameter ratio at midspan of the hole). The aspect ratio depends on the optical characteristics of the beam and on the optical and thermal properties of the material. In deep hole drilling of 0.25 to 1.0 mm in diameter, typical aspect ratios of more than 10:1 have been reported by Yeo et al. (1994). An aspect ratio of 50:1 and drilling tolerance of ±5 to 20 percent of diameter has been reported in the *Machining Data Handbook* (1997).

Taper and barreling. The degree of taper that ranges from 5 to 20 percent of the hole diameter decreases with an increase in the hole depth. The taper depends on the pulse duration and energy, number of pulses, and design of the optical system. In general, the shorter the pulse duration the greater will be the taper. According to Yeo et al. (1994) barreling, formed by the ejection of material from the local sites of the cavity, can be reduced by decreasing the pressure of the assist gas.

Recast layer and microcracking. Selecting a low-power density that effectively expels molten and vaporized material from the machining zone can reduce the recast layer. Using a short pulse duration and a low-repetition rate to minimize heating in the surrounding material can reduce microcracking. Generally, the smaller depth of the altered layer associated with high cutting rates is an advantage of LBM.

Entrance angle. The laser beam is capable of drilling holes with an entrance angle that varies from 10° up to normal incidence at 90° with acceptable hole characteristics. As the entrance angle with respect to the surface becomes smaller, the effective surface power density is reduced. Under such circumstances, it is necessary to increase the power of the laser or improve the quality of the beam.

Drilling time. Deep hole drilling up to 17 mm can be accomplished by operating the laser beam in the rapid-pulsed mode so that the total time is proportional to the amount of energy delivered per pulse. Yeo et al. (1994) found that holes with a similar aspect ratio can be drilled in a similar machining time (an aspect ratio of 20:1 requires 7 s).

TABLE 5.3 Laser Drilling Capabilities

Hole feature	Range
Diameter	0.005–1.27 mm, larger in trepanning
Depth	1.7 mm
Angle	15–90°
Taper	5–20% hole diameter
Depth/diameter ratio	50:1
Drilling depth (trepanned)	6.4 mm
Tolerances	±5–20 % diameter
Minimum corner radious	0.25 mm

Parameters affecting the quality of holes. Table 5.3 lists the laser beam drilling capabilities, while Fig. 5.37 presents the various factors that affect the quality of drilled holes.

Pulse energy. It is recommended that the required peak power should be obtained by increasing the pulse energy while keeping the pulse duration constant. Drilling of holes with longer pulses causes enlargement of the hole entrance.

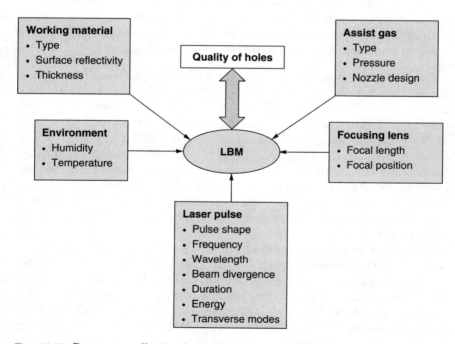

Figure 5.37 Parameters affecting the quality of laser-drilled holes.

Pulse duration. The range of pulse durations suitable for hole drilling is found to be from 0.1 to 2.5 ms. High pulse energy (20 J) and short pulse duration are found suitable for deep hole drilling in aerospace materials.

Assist gases. The gas jet is normally directed with the laser beam into the interaction region to remove the molten material from the machining region and obtain a clean cut. Assist gases also shield the lens from the expelled material by setting up a high-pressure barrier at the nozzle opening. Pure oxygen causes rapid oxidation and exothermic reactions, causing better process efficiency. The selection of air, oxygen, or an inert gas depends on the workpiece material and thickness. An inert gas such as helium or argon should be used if a cut edge of a better quality is required. It should be borne in mind that the smaller the diameter of the gas nozzle and the narrower its distance from the workpiece surface, the better is the quality of the cut. Gas pressure also plays a significant part in the determination of quality and rate of machining.

Material properties and environment. These include the surface characteristics such as reflectivity and absorption coefficient of the bulk material. Additionally, thermal conductivity and diffusivity, density, specific heat, and latent heat are also considered. In this regard, the minimum laser energy required to vaporize plastic is generally lower than that for metals. The removal rate is enhanced for materials of low thermal conductivity, diffusivity, and melting point.

Gagilano et al. (1969) used a pulsed ruby laser for piercing holes in diamond wire drawing dies of 1.4-mm diameter. Two hundred and fifty pulses of energy 2 to 3 J, duration 0.6 ms, and 1 pulse per second were used to perform an opening of 0.46 mm in the die. In case of drilling with multiple pulses, an aspect ratio of 40:1 for a 0.127-mm-diameter hole through 12.5-mm-thick material has been reported by Bellows and Kohls (1983). They also found that the depth of holes produced increases initially as the diameter widens, up to a limit beyond which the penetration depth decreases. They draw attention to small hole-drilling applications such as fuel filters, carburetor nozzles, and jet engine blade cooling holes.

A neodymium-doped yttrium-aluminum-garnet (Nd-YAG) laser, fitted with CNC, on five independent axes for positioning components weighing up to 1 kg, has been used in the aircraft engine industry by Corfe (1983). Accordingly, holes in gas turbine components and nozzle guide vans, 8 mm long and with an aspect ratio of 10:1, were produced, thus realizing savings of 20 to 30 percent over EDM. Figure 5.38 shows laser-drilled cooling holes in a turbine vane.

Figure 5.38 Cooling holes in turbine vane (Corfe, 1983).

Boehme (1983) pointed out that pulsed solid-state lasers produce hole diameters between 0.1 and 0.5 mm at a rate of 0.1 to 10 holes per second, the depth-to-diameter ratio lying in the range 1:1 to 10:1. During laser hole drilling, optical pumping arises from the internal reflections of laser light as the hole becomes progressively deeper. The maximum hole depth achievable is, therefore, diminished by the energy lost due to reflection, from the hole wall, and also by the decrease of the hole aperture. Another detrimental effect to laser machining is the production of plume vapor within the cavity that absorbs laser energy; as a result, a shallower hole is produced. Under such circumstances, repeated pulses are recommended by McGeough (1988) to overcome such process limitations.

The problems of limited maximum hole depth, noncylindrical profile, and the presence of excessive recast material are, normally, associated with conventional lasers. Lau et al. (1994) introduced the ultrasonic-assisted laser machining technique not only to increase the hole depth but also to improve the quality of holes produced in aluminum-based metal matrix composites (MMC). Using such a method, the hole depth was increased by 20 percent in addition to the reduced degree of hole tapering.

5.2.3.2 Cutting and grooving. Powers in the range of 200 W to 1 kW are recommended for cutting through steel. Oxygen-jet assistance is found to be suitable for most metals, as the gas liberates additional energy by the exothermal chemical reactions occurring in the machining region. Boehme (1983) described the use of a 500-W CO_2 laser in conjunction with a CNC system for cutting steel plates of thickness up to 5 mm. Accordingly, the cutting speed has been found to decrease with an increase in the workpiece thickness. CNC laser cutting has also been used in the production of clothes. Accordingly a system incorporating a 400-W laser beam, guided by movable mirrors, is used to cut a 2-m-wide moving fabric at a rate of 80 m/min. The required styles are held in the memory bank of the computer. Cutting of flat quarts of thickness 5 to 100 mm has also been performed by traversing 100 to 250 W continuous output CO_2N_2He laser focused to a diameter of 25 to 100 µm across the material. Furthermore, McGeough (1988) reported that a 300-W laser beam of the same type can be used to cut through 3.1-mm-thick steel plate at 1.02 mm/min, leaving a heat-affected layer of 0.5 mm width. Laser cutting of electrical sheets offered an alternative machining method that avoids the high wear of conventional cutting tools mainly caused by the high silica content (1.4%) of the material. The high flexibility of a CNC laser-workstation in combination with a CAD/CAM system gave the advantage of the prototype and short-run production of sheet blanks of complex geometry such as rotor and stator blanks. Using such a method, laser cutting of electrical sheets became possible at high speeds of 8 m/min when using a 500-W CO_2 laser and a 0.65-mm-thick sheet with an excellent cut quality. A burr height of less than 20 µm at the bottom of the cut has been reported by Tonshof and Dickmann (1989).

Fiber composites are sensitive to conventional cutting and drilling due to their inhomogenous and unisotropic nature. Delaminating, splintering, the presence of burrs, and short tool life caused by the abrasive nature of the fibers are typical problems. In laser cutting of composites, the effect of material anisotropy such as the fiber arrangement is of less importance. Lau et al. (1990) used Nd-YAG laser for cutting carbon fiber composite materials at faster rates than EDM and the cut surface is less heterogeneous. Compressed-air assistance provided a larger depth of cut than argon; however, argon gave smooth and less thermal damage to the cut. They added that, the fiber orientation affected the depth of cut besides the width of the heat-affected zone. Standard (coaxial) oxygen-assisted laser cutting of stainless steel sheets does not provide satisfactory performance. The oxide dross that clings to the bottom edges of the cut forms a hard burr, thus limiting the thickness to be cut by 3 mm. For machining of larger thickness, the machining speed is, consequently, lowered and the finished edge is similar to an oxyacetylene flame cut. The main obstacles in laser machining of stainless steel are due to the low fluidity of the melt and partially the high melting point of chromium oxide (Cr_2O_3) that also restrains oxygen diffusion in the molten cutting front.

In order to solve the problems associated with laser cutting of stainless steel, Hsu and Molian (1995) developed a laser machining technique that employs dual gas jets (shown in Fig. 5.39) to remove the viscous stage in the molten cutting front and, thereby, allowing stainless steel to be cut faster, cleaner, and thicker. For a better understanding of process behavior and to achieve and maintain high-quality machining processes, developments in the area of monitoring, diagnosis, regulations,

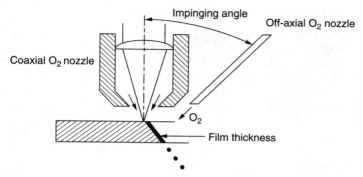

Figure 5.39 Off-axial gas-assisted laser cutting (Hsu and Molian, 1995).

and modeling become essential. In this regard Yilbas (1996) studied the formation of strias by monitoring the high emission, from the upper surface of the workpiece, during laser cutting using fiber-optic cable and a fast-response photo detector.

Recently Ueda et al. (1997) and Spur et al. (1997) investigated the temperature of a workpiece irradiated with a CO_2 pulsed laser in order to determine suitable conditions for cleavage cutting of brittle materials. In an attempt to control the heat conduction in the workpiece material during laser grooving, Chryssolouris et al. (1993) adopted the use of a water jet in tandem with the laser beam thus reducing the heat-affected zone by 70 percent.

5.2.3.3 Texturing and structuring.

Textured steel and aluminum strips or sheets are widely used particularly among car manufacturers. Texturing facilitates lubrication during metal forming and prevents adherence of sheets during annealing. The process is applied to the roll, from which the steel or aluminum sheet takes its surface pattern during the final stages of cold rolling. LBM has replaced the conventional SB method. Accordingly, the product designers can accurately specify the texture putting in mind that the hardness of the roll is not a limitation. In laser beam texturing (LBT), a beam of light from a high-energy gas laser is mechanically chopped into pulses before focusing onto the work roll surface. The high energy causes local melting of the roll, which with gas jet assistance, results in the formation of a discrete crater. The roll is rotated and the laser beam moved axially along the roll in a manner analogous to conventional turning. The depth and size of the crater can be controlled by the energy and dwell time of the beam. Simao and co workers (1994) concluded that LBT can produce regular patterns due to the periodic nature of the process. Figure 5.40 shows typical examples of

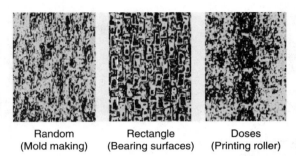

| Random | Rectangle | Doses |
| (Mold making) | (Bearing surfaces) | (Printing roller) |

Figure 5.40 Structuring of surfaces with PM Nd-YAG laser radiation (Tonshof and Sturmer, 1995).

textured surfaces that are widely used in mold making, bearings, and printer roll manufacturing, respectively, as described by Tonshof and Sturmer (1995).

5.2.3.4 Dressing of grinding wheels. Lasers can create well-defined grooves on the grinding wheels due to the evaporation and damage of the composite wheel material. Suitable selection of the dressing feed will result in the dressing of the whole wheel surface thus changing its topography. Laser dressing produces microcutting edges due to the formation of microcraters on the worn-out grains. When these craters are formed in the bond, the grits are loosened and subsequently removed due to an insufficient volume of bond surrounding the grain. Further multiple cracks that are induced thermally during laser dressing help to remove the resolidified layer during grinding, which then exposes new cutting edges. It is also possible to refresh the wheel by evaporating only the metal chips clogged on the wheel by focusing the laser beam suitably; however, selection and control of the laser beam should be carefully considered as mentioned by Ramesh et al. (1989).

5.2.3.5 Milling. Laser milling employing pulsed CO_2 laser has been investigated by Hsu and Copley (1990) in graphite as a model material. In this process, the material is removed by scanning the focused beam across the surface of the workpiece leaving behind a series of narrow, parallel, overlapping grooves. In order to remove layers bounded at the edges by upright walls perpendicular to the layer plane, the laser beam is tilted with respect to the layer plane. Workpiece tilting depends on the incident power, beam speed, groove direction, and optical absorption coefficient of the material. Todd and Copley (1997) developed a prototype laser processing system for shaping advanced ceramic materials.

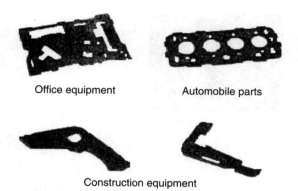

Figure 5.41 CO_2 laser cutting samples.

This prototype is a fully automated, five-axis, closed-loop controlled laser shaping system that accurately and cost effectively produces complex shapes in the above-mentioned material.

5.2.3.6 Fine cutting and drilling. Tabata et al. (1996) reported that CW CO_2 lasers are widely used for fine cutting of steel plates in the automotive, electric, and electronic industries; office equipment; household appliances; industrial machine parts; construction equipment parts; and construction materials. In fine cutting the thickness of the steel plate varies from 0.5 to 16 mm, allowing a cutting speed to vary from 6 to 1.0 m/min, respectively. As depicted in Fig. 5.41, the achieved fine cuts were of narrow width and the cutting surface was clean and smooth. They also added that high-power Nd-YAG lasers with high-beam quality can be used for steel plate cutting in place of CO_2 lasers. The Nd-YAG lasers can also machine nonferrous metals such as copper, brass, and aluminum alloys more efficiently than CO_2 laser since Nd-YAG laser is less reflective to such metals. Plates of thickness from 0.2 to 2 mm are finely cut with the output of 200 W. A cutting speed of 5 m/s for 1-mm-thick plate has been also reported. The slit width and surface cleanliness are almost similar to CO_2 laser processing.

Cao and Huang (1995) invented a new technique for microsize (< 0.07 mm) laser machining, with varying curved wall surfaces diamond wire drawing dies using CW Nd-YAG laser. Repetition frequency, laser power, and feed step of the laser head along the z axis are the most critical parameters. Microhole drilling, in a single pulse, using a pulse slab Nd-YAG laser has been reported by Rohde and Verboven (1995). Accordingly, holes with diameters of 35 µm in a steel plate of 1.2 mm and diameters of 6 µm in 0.03 mm steel foil have been obtained (Fig. 5.42).

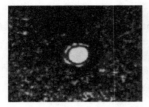 

35 μm hole in 1.2 mm   6 μm in 0.03 mm

Figure 5.42 Fine drilling of holes in steel.

The rectangular spot slab laser beam can also create a slit, using a single pulse. Industrial applications for food industries are sieves with slits of 0.06 mm × 2 mm in stainless steel of 0.5 mm thickness at a rate of 100 slits per second. Several pulses are used to create a hole in percussion drilling mode. A 60-μm-diameter hole has been obtained in 100 $MnCrV_4$ steel plate of 4 mm thickness with an aspect ratio of 65:1 using four to five pulses. Micro blind holes of 30-μm diameter with a maximum aspect ratio 12:1 to 800 μm with aspect ratio 4:1 have also been reported by Rohde and Verboven (1995). It is accordingly recommended to adjust the pulse duration and frequency to evacuate the molten material from the path of the laser beam.

The trepan technique drills holes of diameters larger than the laser beam diameter and offers smaller conicity and higher repeatability of the hole diameter. Using such a method, holes inclined to the surface into fuel-injection nozzles have become possible. Microcutting of 1-mm-thick steel plate leaving a kerf of 85 μm at the entrance side and 60 μm at the exit side and leaving a taper of 1.25 percent have been reported.

5.2.3.7 Marking of computer keyboard. Ricciardi et al. (1996) developed an excimer laser-based technique for marking computer keyboards with high speed and flexibility of the production line that avoids the use of toxic solutions in the marking process. Laser marking does not involve the introduction of any foreign materials or exert any mechanical stresses on the workpiece. The practical advantages of excimer lasers in marking processes are due to the short laser wavelength and pulse duration. High photon energy induces photochemical reactions in the material, resulting in a color change with negligible side effects. The laser marking cell is able to meet the requirements with nominal throughput of one keyboard every 15 s.

5.2.3.8 Trimming of electronic components. Lasers of high power density, small spot size, and short pulse length are useful devices for modification of electronic components, often by selective evaporation. By proper monitoring, the device can be trimmed or adjusted by the required amount. Trimming of resistors composed of chromium-silicon oxide, deposited on the surface of a silicon chip, has been achieved using an argon-ion laser operated at a power density of 0.8 MW/cm^2 with a pulse width of less than 10 µs. Energy from the laser was used to heat the resistor film to about 1000°C thereby altering its resistance value (McGeough, 1988).

5.2.3.9 Controlled fracturing. Laser energy can also be used to fracture or break, in a controlled fashion, delicate items such as electronic circuits or components. The absorption of the beam, focused on a small area of the surface, creates thermal gradients which in turn lead to the formation of mechanical stresses that are sufficient to cause shearing of the material over a region so small that the fracture does not propagate in an uncontrolled manner. Using such a technique, high alumina ceramic, 0.46 mm thick, has been fractured in this way with a 100-W CW CO_2N_2He laser.

5.2.3.10 Dynamic balancing of gyro components. Laser trimming can achieve highly accurate balancing; with material removal at a rate of milligrams per pulse, shallow holes are produced. The dynamic balancing is simplified by utilization of a signal from the points of the imbalance to trigger the laser pulse.

5.2.3.11 Scribing. Brittle materials like silicon, glass, and ceramic may be effectively separated or shaped by laser scribing. With this technique, material is removed by laser along a specific path on the surface of the workpiece. When the material is sufficiently stressed, a fracture occurrs along the scribed path. Silicon transistor wafers have been successfully scribed with a repetitively Q-switched Nd-YAG laser of peak power 300 W, pulsed, at a rate of 400 pulses per second and of duration 300 ns. The rate of scribing reported by McGeough (1988) is 1.5 m/min.

5.2.3.12 Micromachining. Tonshof et al. (1993) investigated the application of excimer laser radiation in a multiaxis CNC micromachining unit with modular design. The unit is used to generate microstructures in ceramic and polymer surfaces. Using such an arrangement, it was possible to produce micromechanical parts of these materials in the fiber-reinforced plastics.

5.2.3.13 Laser-assisted EDM. Allen and Huang (1997) developed a novel combination of machining processes to fabricate small holes. Before the micro-EDM of holes, copper vapor laser radiation was used to obtain an array of small holes first. These holes were then finished by micro-EDM. Their method showed that the machining speed of micro-EDM had been increased and electrode tool wear was markedly reduced while the surface quality remained unchanged.

TABLE 5.4 Laser Beam Selection Guide

Application		Laser type
Drilling	Small holes, 0.25 mm	Ruby, Nd-Glass, Nd-YAG
	Large holes, 1.52 mm	Ruby, Nd-Glass, Nd-YAG
	Large holes, trepanned	Nd-YAG, CO_2
	Drilling, percussion	Ruby, Nd-YAG
Cutting	Thick cutting	CO_2 + gas assistance
	Thin slitting, metals	Nd-YAG
	Thin slitting, plastics	CO_2
	Plastics	CO_2
Materials	Metals	Ruby, Nd-Glass, Nd-YAG
	Organics and nonmetals	Pulsed CO_2
	Ceramics	Pulsed CO_2, Nd-YAG

In general, LBM offers a wide range of machining applications and can tackle different engineering materials. The choice of the type of laser therefore depends on the application required in addition to the material being machined. Table 5.4 provides a laser beam selection guide.

5.2.4 Advantages and limitations

Advantages

- Tool wear and breakage are not encountered.
- Holes can be located accurately by using an optical laser system for alignment.
- Very small holes with a large aspect ratio can be produced.
- A wide variety of hard and difficult-to-machine materials can be tackled.
- Machining is extremely rapid and the setup times are economical.
- Holes can be drilled at difficult entrance angles (10° to the surface).
- Because of its flexibility, the process can be automated easily such as the on-the-fly operation for thin gauge material, which requires one shot to produce a hole.
- The operating cost is low.

Limitations

- High equipment cost.
- Tapers are normally encountered in the direct drilling of holes.
- A blind hole of precise depth is difficult to achieve with a laser.

- The thickness of the material that can be laser drilled is restricted to 50 mm.
- Adherent materials, which are found normally at the exit holes, need to be removed.

5.3 Electron Beam Machining

5.3.1 Introduction

The earliest work of material removal utilizing an electron beam was attributed to Steigerwald who designed a prototype machine in 1947. Electron beam machining (EBM) has been used in industry since the 1960s, initially in nuclear and aerospace welding applications. Drilling small holes, cutting, engraving, and heat treatment are a set of modern applications used in semiconductor manufacturing as well as micromachining areas.

5.3.2 Basic equipment and removal mechanism

The main components of EBM installation, shown in Fig. 5.43 are housed in a vacuum chamber, evacuated to about 10^{-4} torr. The tungsten filament cathode is heated to about 2500 to 3000°C in order to emit electrons. A measure of this effect is the emission current, the magnitude of which varies between 20 and 100 mA. Corresponding current densities lie between 5 and 15 A/cm^2. Emission current depends on the cathode material, temperature, and the high voltage that is usually about 150 kV. Such a high voltage accelerates a stream of electrons in the direction of the workpiece. After acceleration, electrons, focused by the field, travel through a hole in the anode. The electron beam is

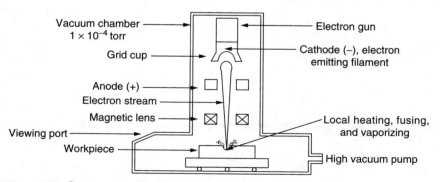

Figure 5.43 Components of an EBM system.

then refocused by a magnetic or electronic lens system so that the beam is directed under control toward the workpiece. The electrons maintain the velocity (228×10^3 km/s) imparted by the acceleration voltage until they strike the workpiece, over a well-defined area, typically 0.25 mm in diameter.

The kinetic energy of the electrons is then rapidly transmitted into heat, causing a corresponding rapid increase in the temperature of the workpiece, to well above its boiling point, thus causing material removal by evaporation. With power densities of 1.55 MW/mm^2 involved in EBM, virtually all engineering materials can be machined by this machining technique. Accurate manipulation of the workpiece coupled with the precise control of the beam is reported by McGeough (1988) to yield a machining process that can be fully automated.

The way in which the focused beam penetrates the workpiece is not completely understood, owing to the complexity of the mechanism involved. However, it is believed that the workpiece surface is melted by a combination of electron pressure and surface tension. The melted liquid is rapidly ejected and vaporized, thus causing material removal rates of about 10 mm^3/min. A pulsed electron beam at 10^4 Hz reduces the temperature of the workpiece outside the region being machined. An early attraction of EBM was the comparatively large depth-to-width ratio (100:1) with applications in fine hole drilling becoming feasible. The absence of mechanical contact and the suitability for automatic control enhance the process capabilities, but the necessity to work in a vacuum lengthens the floor-to-floor cycle time.

The number of pulses required to remove a hole of depth g can be described by

$$n_e = \frac{g}{g_e}$$

The machining time t_m can be given by

$$t_m = \frac{n_e}{f_p}$$

$$f_p = \frac{1}{t_p + t_i}$$

The drilling rate Ψ (mm/min) can therefore be calculated by

$$\Psi = \frac{g f_p}{n_e}$$

According to Kaczmarek (1976), the number of pulses n_e can simply be described as a function of the accelerating voltage V_a and the emission current I_e by

$$n_e = \frac{1}{KI_e V_a}$$

Hence, the drilling rate Ψ (mm/min) and the volumetric removal rate (VRR) become

$$\Psi = Kgf_b I_e V_a$$

$$\text{VRR} = \frac{\pi}{4} Kd_b^2 gf_b I_e V_a$$

In case of slotting a depth g and length L, the slotting time t_m is

$$t_m = \frac{n_e \cdot L}{f_b \cdot d_b}$$

The slotting rate η (mm/min) becomes

$$\eta = Kd_b f_p I_e V_a$$

The VRR (mm³/min) can be calculated by

$$\text{VRR} = Kd_b^2 gf_p I_e V_a$$

where g_e = depth of hole removed per pulse, mm
g = depth of hole or slot required, mm
f_p = frequency of pulses, s⁻¹
t_p = pulse time, μs
t_i = pulse interval, μs
d_b = beam diameter in contact with the workpiece (slot width), mm
V_a = beam accelerating voltage, kV
I_e = beam emission current, mA
K = constant
L = slot length, mm

The depth of penetration depends on the beam diameter, power density, and the accelerating voltage. Moreover the depth of eroded material per pulse depends on the density of the workpiece material as well as on the beam diameter. Practically, the number of pulses that produce a given hole depth is usually found to decrease with an increase in the acceleration voltage. For a fixed set of process conditions, the number

160   Chapter Five

of pulses required increases hyperbolically as the depth of the hole increases. In practical terms, this conclusion means that when a certain depth has been reached, any further EBM to deepen the hole would require a very large increase in the number of pulses. The machining time, in EBM, required to drill a hole depends on the number of pulses required to erode a certain depth and pulse frequency. For slotting by EBM, the machining time is affected by slot length, beam diameter, pulse duration, and number of pulses required to remove a specified depth. The EBM rate is usually evaluated in terms of the number of pulses required to evaporate a particular amount of material. The use of electron counters, which register the number of pulses, enables ready adjustment of the machining time to produce a required depth of cut. Workpiece material properties such as boiling point and thermal conductivity play a significant role in determining how readily they can be machined. Other thermal properties such as electrical conductivity are considered as additional factors. Figure 5.44 summarizes the factors affecting the performance of EBM.

Fig. 5.45 shows the number of pulses versus the relative machinability index. In this case, cadmium is considered as the best machinable material since it requires the minimum number of pulses. A further index utilizes the relative power required to remove an equal volume of various metals in equal machining time based on aluminum is shown in Fig. 5.46. Materials having a lower power consumption utilize a

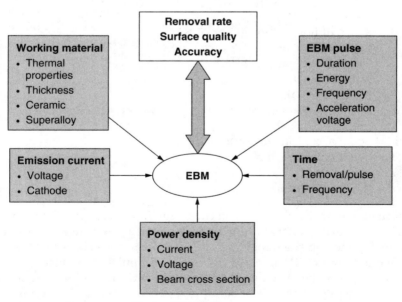

Figure 5.44 Parameters affecting EBM performance.

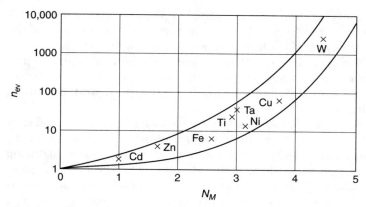

Figure 5.45 The number of pulses required (vertical) versus the relative machinability index (Kaczmarek, 1976).

smaller number of pulses to remove the same volume, and hence the machinability is improved. The machinability is affected jointly by the thermal properties of the material as well as by the density and electrical conductivity. Studies have shown that increasing the accelerating voltage above 120 kV is quite ineffective. The number of pulses, taken as a machinability index, depends on the hole sinking depth. Increasing the pulse duration reduces the number of pulses required due to the higher energy available in this case. The diameter of holes or cavities obtained as a result of pulses depends on the material being machined as well as on the pulse energy. The maximum groove width increases linearly with the pulse charge. For a given pulse charge, Ag has the largest groove width that constitutes a large removed volume and hence the highest machinability.

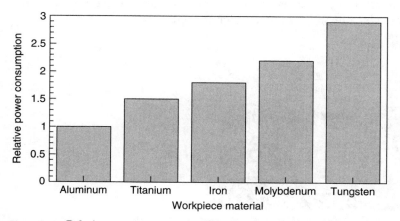

Figure 5.46 Relative power consumption for different materials (El-Hofy, 1995).

Practically, cavity diameters in the range of 35 to 40 µm are obtainable with out-of-roundness from 3 to 6 µm in more than 50 percent of the cases. The diameter of the altered layer is proportional to the hole diameter which is reported to be 40 to 75 µm for a hole of 20 to 55 µm. Conditions leading to larger removal rates and high machinability are reported to be accompanied by a larger layer thickness and, moreover, a greater surface roughness.

The effect of pulse numbers on the accelerating voltage showed that increasing the hole depth requires a much greater rise in the number at low voltage, due mainly to a relative rise in heat losses resulting from the conduction and melting of the adjacent metal layers. For a given number of pulses, little improvement in material removal rate is obtained by increasing the accelerating voltage above 120 kV. The increase of pulse duration raises the pulse energy available, which in turn reduces the number of pulses required to obtain the required machining result. Kaczmarek (1976) quoted an optimum working distance at which a minimum number of pulses are required. He pointed out that a focal point just below the upper surface of the workpiece is sometimes the most effective. The drilling rate by EBM (in holes per second) decreases with an increase in the thickness of the workpiece as well as in the diameter of the hole to be produced.

The surface quality produced by EBM depends on the type of material. In this regard, the surface roughness increases with pulse charge for nickel, carbon, gold, and tungsten. Estimates of surface roughness for small holes and cuts are near to 1 µm R_a. Surface layers of material treated by EBM are affected by the temperature of the focused beam, illustrated by the white layer ring surrounding the hole, shown in Fig. 5.47. The diameter of the damaged layer increases with pulse duration and hole diameter. A typical heat-affected zone can be as much as 0.25 mm in EBM, which can be detrimental to the structural integrity of highly stressed components and, for such components, should be removed.

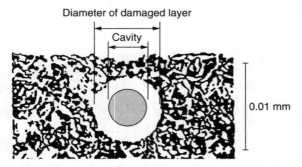

Figure 5.47 Cross section of cavity by single EBM pulse in chromium molybdenum steel (Kaczmarek, 1976).

5.3.3 Applications

5.3.3.1 Drilling. Steigerwald and Mayer (1967) considered that for successful application, improved reproducibility, greater working speeds, and deeper holes of accurately controlled shapes are all needed. Boehme (1983) discussed drilling applications with an electron beam machine fitted with a system for numerically controlling the beam power, focus and pulse duration, and mechanical motion. Cylindrical, conical, and barrel-shaped holes of various diameters can be drilled with consistent accuracy at rates of several thousand holes per second. Holes at an inclination angle of about 15° were also possible. Boehme (1983) claimed that the largest diameter and depth of holes that can be accurately drilled by EBM are, respectively, 1.5 mm and 10 mm and that the aspect depth-to-diameter ratio is normally in the range of 1:1 to 1:15. For deeper holes, in the range of 2.5 to 7.5 mm, Steigerwald and Mayer (1967) emphasized the need for a stable power supply that can emit the required groups of pulses and that, for a well-controlled beam of closely defined diameter, the angle of aperture has a strong bearing on the shape of the hole produced. Holes of about 19 mm were produced. Drew (1976) showed EB-drilled holes in a superalloy turbine blade at angles of 60° to 90°.

5.3.3.2 Perforation of thin sheets. For perforation by EBM to be economically acceptable, 10^4 to 10^5 holes per second have to be produced. Thus single pulses lasting only a few microseconds are needed. In some applications the sheet or foil is stretched on a rotating drum, which is simultaneously shifted in the direction of its axis. Rows of perforations following a helical line are thereby produced. Manipulators capable of linear and rotating movements in four axes are used for EBM perforation of jet engine components. Foil made of a synthetic material has been perforated with 620 holes per square millimeter for filter application at a rate of one hole every 10 µs. EBM perforation can be applied to the production of filters and masks of color television tubes. Other applications for perforation lie in sieve manufacture, for sound insulation and in glass fiber production.

5.3.3.3 Slotting. Rectangular slots of 0.2 by 6.35 mm in 1.57-mm-thick stainless steel plate are produced in 5 min using 140 kV, 120 µA, a pulse width of 80 µs, and a frequency of 50 Hz. The rate of slotting depends on the workpiece thickness. In this regard 0.05-mm-thick stainless steel was cut at a rate of 100 m/min, while 0.18-mm-thick stainless steel was cut at 50 m/min using similar machining conditions.

5.3.3.4 Integrated circuit fabrication. Birnie and Champney (1967) drew the attention to use of electron beam technology in scribing thin film circuits for the electronic industry. A detailed account for the manufacture of integrated circuits has been presented by Yew et al. (1994). The beam is positioned accurately by means of deflection coils at the location where a pattern is to be written, by exposing a film of electron resist coated on either a chrome mask blank or wafer, for the production of the lithographic definition required. An electron beam of energy about 10 to 20 kV can either break the bonds between the polymer molecules or cause cross-linking in the polymers.

With the onset of either of these conditions, the solubility changes when the resist film is immersed in the developer, usually a solvent for the resist. Because of the difference in solubility between the original and exposed resist polymers, a differential material removal occurs. A fine pattern of polymer is thus obtained, which is used as an active mask to avoid unwanted etching of the integrated circuit mask. Pattern generation can be carried out by a vector scan. Accordingly, the electron beam is deflected only to the locations at which the electron resist is to be exposed. As soon as the deflection system completes the positioning of the beam, at the required place, the electron beam action is started. With the raster scan method, described by McGeough (1988), the chip pattern is divided into subfields. The electron beam in a raster, like that employed with the television, scans each subfield. The electron beam is turned on and off along each raster line as needed. The required pattern is fully formed by the combined effects of electron beam exposure and subsequent resist development. Fig. 5.48 shows a typical application where EBM is used for a miniature electronic hybrid circuit engravement with a 40-μm-wide trace. Direct wafer processing by EBM is necessary for feature sizes below one or two microns. Line width as small as several hundredth of an angstrom can also be written with electron beam techniques at speeds as high as 20 MHz. Direct wafer processing electron beam systems are capable of producing 22 wafers per hour.

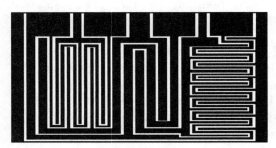

Figure 5.48 Hybrid circuit engraved with 40 μm traces at speeds > 5 m/s.

TABLE 5.5 EBM Process Parameters and Capabilities

EBM parameter	Level
Acceleration voltage	50–60 kV
Beam current	100–100 µA
Beam power	0.5-50 kW
Pulse time	4-64,000 µs
Pulse frequency	0.1-16,000 Hz
Vacuum	0.01-0.0001 mm mercury
Spot size	0.013-0.025 mm
Deflection range	6.4 mm^2
Beam intensity	1.55×10^5–1.55×10^9 W/cm^2
Depth of cut	Up to 6.4 mm
Narrowest cut	0.025 mm in 0.025-mm-thick metal
Hole range	0.025 mm in 0.02-mm-thick metal
	1.0 mm in 5-mm-thick metal
Hole taper	1°–2° typical
Hole angle to surface	20°–90°
Removal rate	40 mm^3/s
Penetration rate	0.25 mm/s
Perforation rate	Up to 5000 holes/s
Tolerance	± 10% depth of cut
Surface roughness	1 µm R_a

Electron beams can also be used to engrave ceramic as well as vapor-deposited metallic layers on substrate. EBM, process parameters and general capabilities are shown in Table 5.5.

5.3.4 Advantages and disadvantages

Advantages

- Drilling is possible at high rates (up to 4000 holes per second).
- No difficulty is encountered with acute angles.
- Drilling parameters can easily be changed during machining.
- No limitation is imposed by workpiece hardness, ductility, and surface reflectivity.
- No mechanical distortion occurs to the workpiece since there is no contact.
- The process is capable of achieving high accuracy and repeatability of 0.1 mm for position of holes and 5 percent for the hole diameter.
- The process produces the best surface finish compared to other processes.
- The cost is relatively small compared to other processes used to produce very small holes.

Disadvantages

- High capital equipment cost
- Long production time due to the time needed to generate a vacuum
- The presence of a thin recast layer
- Need for auxiliary backing material

5.4 Plasma Beam Machining

5.4.1 Introduction

When the temperature of a gas is raised to about 2000°C, the gas molecules become dissociated into separate atoms. At higher temperatures, 30,000°C, these atoms become ionized. The gas in this stage is termed *plasma*. Machining by plasma was adopted in the early 1950s as an alternative method for oxy-gas flame cutting of stainless steel, aluminum, and other nonferrous metals. During that time the process limitations regarding the low cutting speed, poor machining quality, and the unreliable equipment were clear. Recently machining of both metallic and nonconductive materials has become much more attractive. An important feature of plasma beam machining (PBM), is that it is the only fabricating method that works faster in stainless steel than it does in mild steel.

5.4.2 Machining systems

In plasma machining a continuous arc is generated between a hot tungsten cathode and the water-cooled copper anode. A gas is introduced around the cathode and flows through the anode. The temperature, in the narrow orifice around the cathode, reaches 28,000°C, which is enough to produce a high-temperature plasma arc. Under these conditions, the metal being machined is very rapidly melted and vaporized. The stream of ionized gases flushes away the machining debris as a fine spray creating flow lines on the machined surface. The general characteristics of PBM are shown in Table 5.6. The removal rates by this method are

TABLE 5.6 PBM Characteristics

Parameter	Level
Velocity of plasma jet	500 m/s
Material removal rate	150 cm^3/min
Specific energy	100 W/(cm$^3 \cdot$ min)
Power range	2–200 kW
Voltage	30–250 V
Current	Up to 600 A
Machining speed	0.1–7.5 m/min
Maximum plate thickness	200 mm

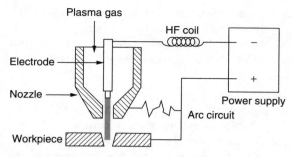

Figure 5.49 Transferred plasma arc system.

substantially higher than those of conventional single-point turning operation. Plasma machining systems are divided into plasma arc, plasma jet, shielded plasma, and air plasma.

5.4.2.1 Plasma arc. As shown in Fig. 5.49, the arc is struck from the rear electrode of the plasma torch to the conductive workpiece causing temperatures as high as 33,300°C. The double arcing effect between the nozzle and the workpiece damages the electrode and the workpiece. High heat transfer rates are found to occur during plasma arc due to the transfer of all the anode heat to the workpiece. Owing to the greater efficiency of plasma arc systems, they are often used for machining metals. Plasma arc does not depend on a chemical reaction between the gas and the work metal. Because the temperature is high, the process is suitable for any electrically conductive material including those that are resistant to oxy-fuel gas cutting.

5.4.2.2 Plasma jet. In this system, shown in Fig. 5.50, the nontransferred arc is operated within the torch itself. Only ionized gas (plasma) is emitted as a jet causing temperature as high as 16,600°C. Since the

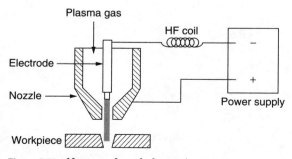

Figure 5.50 Nontransferred plasma jet system.

torch itself is the anode, a large part of the anode heat is extracted by the cooling water and is not effectively used in the material removal process. Nonconductive materials that are difficult to machine, by conventional methods, are often successfully tackled by the plasma jet system.

5.4.2.3 Shielded plasma

Gas-shielded plasma. When machining different materials such as aluminum, stainless steel, and mild steel, assisting gases may have to be used in order to produce cuts of acceptable quality. In such a case an outer shield of gas, is added, around the nozzle, to reduce the effect of the atmosphere on the machining gas (nitrogen or argon). The shielding gas depends on the metal being machined. For stainless steel, aluminum and other nonferrous metals, hydrogen is often used as a shielding gas. Carbon dioxide is popular for ferrous and nonferrous metals. For mild steels, air or oxygen may be also used.

Water-shielded plasma. As shown in Fig. 5.51, nitrogen is used for machining while the shield gas is replaced by water. Water forms a radial jacket around the plasma torch. The cooling effect of water is reported to reduce the width of the cutting zone and improve the quality of cut. However, no improvement in both the cutting rate and the squareness of the cut has been reported by McGeough (1988).

5.4.2.4 Air plasma.
Compressed air is used as the machining gas instead of nitrogen or argon (Fig. 5.52). When air is subjected to the high temperature of the electric arc, it breaks down into its constituent gases. Since the oxygen, in the resulting plasma, is very reactive especially with ferrous metals, machining rates are raised by 25 percent. The main drawback of this method is the heavily oxidized surface, which is frequently obtained in case of stainless steel and aluminum. Because tungsten is reactive with oxygen, hafnium copper (Hf–Cu) or hafnium-zirconium (Hf–Zr) alloys also replace tungsten electrodes. However, the

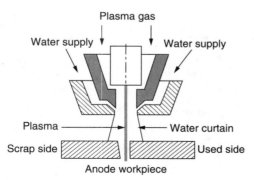

Figure 5.51 Water-shielded plasma.

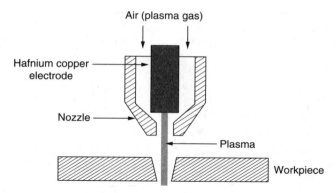

Figure 5.52 Air plasma.

life of these electrodes becomes shorter. Electrically conductive materials such as stainless and chrome nickel alloy steels, aluminum, and copper can be machined by air plasma. Since air is used for machining and shielding purposes, the machining cost is about half that of gas- or water-shielded plasma. When a gas mixture of 80% nitrogen and 20% oxygen is used, the machining rate of mild steel is increased by about 25 percent.

5.4.3 Material removal rate

During PBM absorbing the heat energy from the plasma jet directed to the workpiece activates metal removal. The plasma torch blows the molten and evaporated metal away as a fine spray or vapor. The resulting cutting rates and hence the machinability depend on the workpiece being machined as well as the type of the cutting and shielding gases that determine the maximum temperature transfer rates. The maximum machining speed, as an index of machinability for dual gas plasma of carbon steel, stainless steel, and aluminum, is shown in Fig. 5.53. Figure 5.54 shows the power consumption factor needed in plasma beam rough turning of some alloys. A low factor indicates either low energy required or high removal rates. The machining speed is found to decrease with increasing the thickness of the metal or the cut width in case of beveling. As the power is increased, the efficient removal of melted metal is found to need a corresponding rise in the gas flow rate. During plasma machining of 12-mm-thick steel plate using 220 kW the machining speed is 2500 mm/min, which is 5 times greater than that for oxy-gas cutting.

5.4.4 Accuracy and surface quality

The edges of the workpieces cut by PBM are often beveled. McGeough (1988) reported that the right side of the plasma arc relative to the

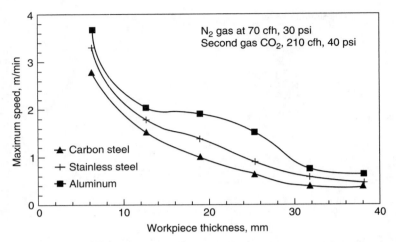

Figure 5.53 Machinability of different material in gas-shielded plasma (El-Hofy, 1995).

cutting direction produces a square edge to within ±3°. The left-hand edge is beveled to about 15° due to the clockwise swirling of the machining gas. Owing to the high rate of heat transfer the depth of fused metal extends to about 0.18 mm below the cut surface. The high machining speed does not allow the heat to penetrate more than a few microns from

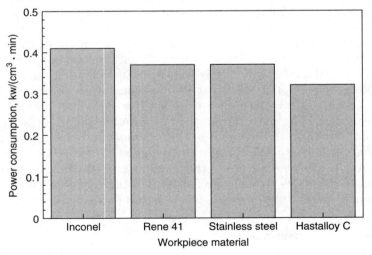

Figure 5.54 Power consumption in plasma beam rough turning (El-Hofy, 1995).

the edges of the cut which produces little or no distortion in the cut workpiece. The cut edge of the material tends to be harder than the base material. A further heat-affected zone (HAZ) of thickness 0.25 to 1.12 mm has been reported. Additionally due to the rapid cooling, cracks may arise beyond the heat-affected zone to 1.6 mm. A clean, smooth surface is produced by PBM. Large tolerances of ±1.6 mm can be achieved. Finish cuts are, therefore, required when narrow tolerances are required.

5.4.5 Applications

1. PAM is an attractive turning method for difficult-to-machine materials by conventional methods. In this regard, cutting speeds of 2 m/min and a feed rate of 5 mm per revolution produced a surface finish of 0.5 mm R_t. The depth of cut can be controlled through the machining power or surface speed (Fig. 5.55).

2. Computer numerical controlled PBM is used for profile cutting of metals that are difficult to tackle by oxyacetylene gas technique such as stainless steel and aluminum. A large number of parts can also be produced from one large sheet thus eliminating shearing operations.

3. PBM can cut 1.5-mm-deep, 12.5-mm-wide grooves in stainless steel at 80 mm^3/min, using 50 kW as the cutting power. Such a high machining rate is 10 times the rate of grinding and chipping methods. Lower machining rates are obtainable when these grooves are cut in nonconductive materials. The groove dimension however depends on the traverse speed, arc power, and the angle and height of the plasma arc.

4. The process is recommended for parts that have subsequent welding operations.

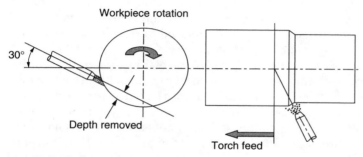

Figure 5.55 Plasma arc turning.

5. A plasma arc can cut tubes of wall thickness of up to 50 mm. In this case no deburring is required before tube welding (*Machining Handbook*, 1989).

6. Underwater NC plasma cutting can achieve machining accuracy of ±0.2 mm in 9 m at low cutting speeds.

5.4.6 Advantages and disadvantages

Advantages. Many advantages of plasma technology have been mentioned at *www.advancedplasma.com/*, including

- Requires no complicated chemical analysis or maintenance
- Uses no harmful chlorinated fluorocarbons, solvents, or acid cleaning chemicals
- Operates cleanly, often eliminating the need for vapor degreasing, solvent wiping, ultrasonic cleaning, and grit blasting
- Requires no worker exposure to harmful chemicals
- Needs less energy to operate

Disadvantages

- The large power supplies needed (220 kW) are required to cut through 12-mm-thick mild steel plate at 2.5 m/min.
- The process also produces heat that could spoil the workpiece and produce toxic fumes.

5.5 Ion Beam Machining

5.5.1 Introduction

Ion beam machining (IBM) takes place in a vacuum chamber using charged ions fired from an ion source toward the workpiece by means of an accelerating voltage. The mechanism of material removal in IBM differs from that of EBM. It is closely related to the ejection of atoms, from the surface, by other ionized atoms (ions) that bombard the work material. The process is, therefore, called ion etching, ion milling, or ion polishing. The machining system, shown in Fig. 5.56, has an ion source that produces a sufficiently intense beam, with an acceptable spread in its energy for the removal of atoms from the workpiece surface by impingement of ions. A heated tungsten filament acts as the cathode, from which electrons are accelerated by means of high voltage (1 kV) toward the anode. During the passage of these electrons from the cathode toward

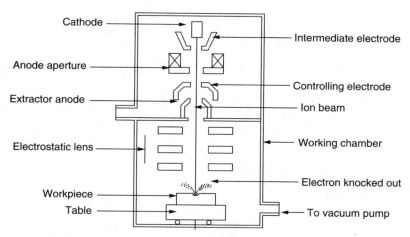

Figure 5.56 IBM system components.

the anode, they interact with argon atoms in the plasma source, to produce argon ions.

$$Ar + e^- \rightarrow Ar^+ + 2e$$

A magnetic field is produced between the cathode and anode that makes the electrons spiral. The path length of the electrons is, therefore, increased through the argon gas, which, in turn, increases the ionization process. The produced ions are then extracted from the plasma toward the workpiece, which is mounted on a water-cooled table having a tilting angle of 0° to 80°. Machining variables such as acceleration voltage, flux, and angle of incidence are independently controlled.

5.5.2 Material removal rate

As shown in Fig. 5.57, if the ions strike the machined surface obliquely, atom ejection is likely to occur from the primary collision. Under such conditions the incident momentum vector is reported to have the greatest influence on the ejection process (McGeough, 1988). The sputtering yield, that is, the number of atoms yielded per incident ion, may be one order of magnitude greater for oblique cutting than normal incidence. The material is, therefore, removed by the transfer of momentum from the incident ions to atoms on the surface of the material. The atom, removed from the surface, is also deflected away from the material. Energies greater than the binding energy of 5 to 10 eV are needed to effect removal of atoms. At higher energies sufficient momentum may

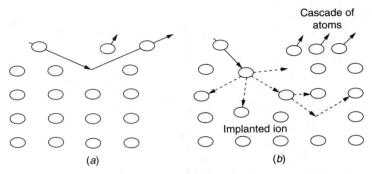

Figure 5.57 Mechanism of material removal at (a) low and (b) high energy levels (McGeough, 1988).

cause the removal of several atoms from the surface (cascade-type effect). Furthermore, the incident ion will become implemented deeper into the material, damaging it, by displacement of atoms. In IBM material is removed by the transfer of momentum from the incident ions to the atoms, in the surface of the material. As a result an atom is removed from the surface, while the ion is deflected away from the material. The amount of yield and hence the machining rate depend on material being machined, the type of ions and their energy, the angle of incidence and, in some cases, the gas pressure as described by McGeough (1988) as

$$V(\theta) = \frac{(9.6 \times 10^{25}) S(\theta) \cos(\theta)}{n}$$

where $V(\theta)$ = etch rate, atoms per min/(mA·cm^{-2})
n = density of target material, atoms per cm^3
$S(\theta)$ = yield, atoms per ion

The $\cos(\theta)$ term takes into account the reduced current densities at angles away from normal incidence. The yield and hence the machining rate depend on the binding energy of atoms in the material being machined. The amount of yield varies with the introduction of gases, which react with the surface of the material, varying its binding energy and hence the rate of material removal. The amount of yield and hence the machining rate present suitable indices of machinability in IBM. Figure 5.58 and 5.59 arrange some materials in descending order with respect to their machinability index. Metallic materials seem to have higher machinability than nonmetallic ones.

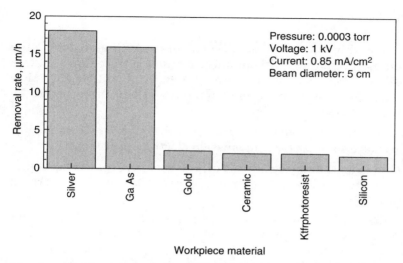

Figure 5.58 Removal rates for different materials (El-Hofy, 1995).

5.5.3 Accuracy and surface effects

Small dimensions as 10 to 100 nm are possible using IBM. The slope of the sidewalls of the machined surface and its surface finish are determined by the angle of incidence of the ion beam. Accuracy levels of ±1.0 percent, with a repeatability of ±1 percent have been reported by McGeough (1988). Surface texturing produces a cone-and-ridge-like configuration on the order of 1 μm in size. However, smoothing to a surface finish less than 1 μm can be obtained.

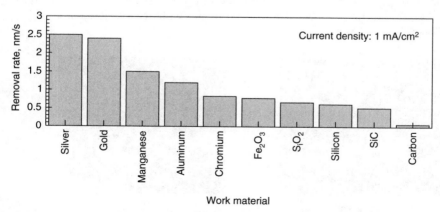

Figure 5.59 Variation removal rate with work material in IBM (El-Hofy, 1995).

5.5.4 Applications

1. IBM is used in smoothing of laser mirrors as well as reducing the thickness of thin films without affecting their surface finish. In this regard thinning of samples of silicon to a thickness of 10 to 15 μm has been reported using argon ions impinging at normal incidence by McGeough (1988).
2. Using two opposing beams, a thin circular region on a rotating sample can produce samples for transmission electron microscopy.
3. Polishing and shaping of optical surfaces by direct sputtering of preforms in glass, silica, and diamond is performed using patterning masks.
4. The process can produce closely packed textured cones in different materials including copper, nickel, stainless steel, silver, and gold (Fig. 5.60). Sputter etching can also create microscopic surface texture using the sputter deposition of a lower yield material on the surface.
5. Atomically clean surfaces can be produced by IBM that are used in the adhesion of gold films to silicon and aluminum oxide substrate. Higher ion energies can be used to remove a layer of the surface oxide.
6. IBM can mill a line width of 0.2 μm, which is used in the fabrication of bubble memory devices of depth-to-width ratios of 2:1 (McGeough, 1988). Problems associated with the alternative chemical etching method such as the lack of undercutting are avoided since masking is only needed to shadow the beam.
7. Further applications regarding two-dimensional polymer base diffraction grating by IBM can be seen in the website *www.blkbox.com/* while ion beam deposition (IBD) in the website *www.skion.com/*.

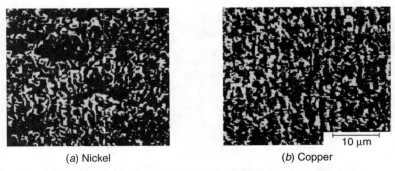

(a) Nickel   (b) Copper

Figure 5.60 Scanning electron photomicrographs (McGeough, 1988).

References

Ahmed, M. S., and Knight, J. A. (1988). "Roll texturing by EDT." *Mechanical Working and Steel Processing Conf. Proc.*, pp. 405–413.

Allen, D., and Lecheheb, A. (1996). "Micro Electrodischarge Machining of Ink Jet Nozzle, Optimum Selection of Material and Machining Parameters," *Journal of Materials Processing Technology*, 58:53–66.

Allen, D., and Huang, S. X. (1995). "An Investigation into Multi-electrode EDM of Micro Holes," *ISEM-11*, pp. 389–392.

Allen, D., and Huang, S. X. (1997). "The Reduction of Tool Wear and Machining Time for Micro-EDM of Micro Holes by Using Copper Vapour Laser Machining as a Roughing Process," *International Journal of Electrical Machining* (IJEM), Tokyo, 2:9–11.

Amalnik, M., McGeough, J., and El-Hofy, H. (1997). "An Intelligent Knowledge Based System for Manufacturability Evaluation of Design for EDT," *32nd MATADOR Conf.*, Manchester, pp. 418–424.

Aspinwal, D. K., Wise, M. L. H., Stout, K. J., Goh, T. H. A., Zhao, F. L., and El-Menshawy, M. F. (1991). "Electrical Discharge Texturing," *Int. J. of Machine Tools and Manufacturing*, 32:183–191.

Aspinwal, D. K., Zhao, F.L., El-Menshawy, M. F. (1989). "Electrical Discharge Texturing (EDT) of Steel Rolls," *Surface Topography*, 2:123–141.

Bellows, G., and Kohls, J. B. (1983). "Drilling without Drill," *American Mechanist*, Special Report 743, pp. 183–185.

Benhadad, M. A., McGeough, J. A., and Barker, M. B. (1991). "Electrodischarge Machining of Al-Li Alloys and Its Effect on Surface Roughness, Hardness and Tensile Strength," *Processing of Advanced Materials*, VI(314):123–123.

Birnie, J., and Champney, M. (1967). "The Contribution of Electron Beams to Machining and Forming in Electrical Methods of Machining and Forming," *IEE Conf.*, Publ. No. 36, pp. 210–221.

Boehme, D. (1983). "Perforation Welding and Surface Treatment with Electron and Laser Beam," *ISEM-7*, pp. 190–199.

Chen, X. K., and McGeough, J. (1994). "Fast Drilling of Small Deep Holes by EDM," *Tenth Int. Conf. on Computer-Aided Production Engineering* (CAPE-10), Palermo, pp. 188–195.

Chryssolouris, G., Sheng, P., and Anastasia, N. (1993) "Laser Grooving of Composite Materials with the Aid of Water Jet," *Trans. of ASME, Journal of Engineering for Industry*, 115:62–72.

Corfe, A. (1983). "Why a Laser Is Better Than EDM for Drilling," *The Production Engineer*, 62(110):13–14.

Crookall, J. R., and Khor, B. C. (1974). "Electro Discharge Machined Surface," *15th MTDR Conf.*, pp. 373–334.

Dauw, D. F., and Van Coppenolle, B. (1995). "On the Evolution of EDM Research," *ISEM-11*. Lausanne Switzerland, pp. 117–142.

Dauw, D. F., and Brown, C. A. (1990). "Surface Topography Investigations by Fractal Analysis of Spark Eroded Electrically Conductive Ceramics," *Annals of CIRP*, 39(1):161–165.

Delpreti, M. R. (1977). "Physical and Chemical Characteristics of Super Facial Layers," *Proc. of ISEM-5*, pp. 209–212.

Devries, M., Duffie, N., Kruth, J., Dauw, D., and Shumacher, B. (1990). "Integration of EDM within a CIM Environment," *Annals of CIRP*, 39(2):1–8.

Drew, J. (1976). "Electron Beam Tackle Tough Machining Jobs," *Machine Design*, 26:94–98.

El-Hofy, H. (1995). "Machinability Indices for Some Non-conventional Machining Processes," *Alexandria Engineering Journal* (AEJ), 34(3):231–245.

El-Hofy, H., De Silva, A., and McGeough, J. A. (1997). "Technological Advances in EDM," *PEDAC-6 Conf.*, Alexandria, pp. 941–954.

El-Menshawy, M. F., and Ahmed, M. S. (1985). "Monitoring and Control of Electro-discharge Texturing Process for Steel Cold Mill Work Rolls," *NAMRIC-13*, Berkeley, USA, pp. 470–475.

Fengguo, C., and Jianyu, H. (1995). "Technique of Precise Laser Machining for Diamond Wire Drawing Dies," *ISEM-11 Conf.*, pp. 629–637.

Fujino, M, Okamoto, N., and Masuzawa, T. (1995). "Development of Multipurpose Micro Processing Machine," *ISEM-11*, pp. 613–620.

Fukuzawa, Y., Tani, T., Ito, Y., Ichinose, Y., and Mohri, N. (1995). "EDM of Insulator Ceramics with a Sheet of Metal Mesh," *ISEM-11*, pp. 173–179.
Gagilano, F. P., et al. (1969). "Lasers in Industry," *IEEE*, 57(2):114–147.
Hsu, M. J., and Molian, P. A. (1995). "Off-axial Gas Jet Assisted Laser Cutting of 6.35 mm Thick Stainless Steel," ASME, *Journal of Engineering for Industry*, 117:272–276.
Hsu, R. K., and Copley, S. M. (1990). "Producing Three-dimensional Shapes by Laser Milling," *Trans. ASME, Journal of Engineering for Industry*, 112:375–379.
Jeswani, M. L. (1977). "Study of Surface Finish in EDM," *ME Journal*, 57:329–333.
Jeswani, M. L. (1978). "Roughness and Wear Characteristics of Spark Eroded Surfaces," *Wear*, 51:227–236.
Kaczmarek, J. (1976). *Principles of Machining by Cutting, Abrasion and Erosion*. Stevenage, U.K.: Peter Pergrenius Ltd.
Kalpakjian, S. (1997). *Manufacturing Processes for Engineering Materials*, 3rd ed. Reading, MA: Addison Wesley.
Kuneida, M., and Furuoya, S. (1991). "Improvements of EDM Efficiency by Supplying Oxygen Gas into Gap," *Annals of CIRP*, 40(1):215–218.
Kozak, J., Dabrowski, L., Rozenek, M., and Cepiak, J. (1995). "Wire EDM of Multi Layer Metal Matrix Composite and Laminate Structures," *ISEM-11*, pp. 427–436.
Lau, W. S., Lee, W. P., and Pans, S. Q. (1990). "Pulsed Nd: YAG Laser Cutting of Carbon Fiber Composite Materials," *Annals of CIRP*, 39(1):179–182.
Lau, W. S., Yue, T. M., and Wang, M. (1994). "Computer Aided Laser Drilling of Aluminium-Based Metal Matrix Composites," *Annals of CIRP*, 43(1):177–180.
Lauwers, B., and Kruth, J. (1995). "A Computer Aided Process Planning System for Electrical Discharge Machining," *ISEM-11*, Lausanne, Switzerland, pp. 261–278 (JSEM/CIRP).
Levy, G. N. (1993). "Environmentally Friendly and High-capacity Dielectric Regeneration for Wire EDM," *Annals of CIRP*, 42(1):227–230.
Levy, G. N., and Maggi, F. (1990). "WED Machinability Comparison of Different Steel Grades," *Annals of CIRP*, 39(1):183–186.
Machining Data Handbook. (1997). 3rd ed. Cincinnati, OH: Machinability Data Center, Institute of Advanced Manufacturing.
Masuzawa, T., and Saki, S. (1978). "Quick Finishing of WEDM Products Using Mate-electrode," *Annals of CIRP*, 36(1):123–326.
Masuzawa, T., Kuo, C. L., and Fujino, M. (1994). "A Combined Electrical Machining Process for Micro Nozzle Fabrication," *Annals of CIRP*, 43(1):189–192.
Masuzawa, T. (1990). "Improved Jet Flushing for EDM," *Annals of CIRP*, 41(1):197–199.
Masuzawa, T., and Tonshof, H. (1997). "Three Dimensional Micro Machining by m/c Tools, *Annals of CIRP*, 46(2):621–628.
McGeough, J. A., and De Silva, A. (1996). "Research Directions and Applications of Unconventional Machining," *Proc. of Advanced Manufacturing Processes, Systems & Technologies* (AMPST), pp. 3–15.
McGeough, J. A. (1988). *Advanced Methods of Machining*, London, New York: Chapman and Hall.
McGeough, J. A. (2002). *Micromachining of Engineering Materials*, New York: Marcel Dekker.
McGeough, J. A., and Rasmussen, H. (1992). "A Model for the Surface Texturing of Steel Rolls by Electro Discharge Machining," *NAMRIC Conf.*, pp. 155–164.
Metals Handbook. (1989). Vol. 16, *Machining*, Meterials Park, OH: ASM International.
Mohri, N., and Saito, M. (1991). "A New Process of Finish Machining on Free Surface by EDM Methods," *Annals of CIRP*, 40(1):207–230.
Morita, A., Imai, Y., Maruyama, H., and Kobyashi, K. (1997). "Fuzzy Controller in EDM," *ISEM-9 Conf.* (JESM/CIRP), Nagoya, Japan, pp. 236–240.
Motoki, M., and Lee, C. (1968). "Electro Spark Machining Using Trapezoidal Power Supply," *Electrochemical Eng. in Japan*, 8(3):12–20.
Pawleski, O., Rasp, W., Zwick, W., Nettelbeck, H. J., and Steinhoff, K. (1994). "The Influence of Different Work Roll Texturing Systems on the Development of Surface Structure in the Temper Rolling Process of Steel Sheet Used in Automotive Industry," *Journal of Material Processing Technology*, 45:215–222.

Ramesh Babu, N., Radhakrishnan, V., and Murti, Y. V. (1989). "Investigation on Laser Dressing of Grinding Wheels—Preliminary Study," ASME, *Journal of Engineering for Industry*, 111:244–252.

Ricciardi, G., Gantello, M., and Savant-Aira, G. (1996). "Marking of Computer Keyboards by Means of Excimer Lasers," *Annals of CIRP*, 45(1):191–196.

Rohde, H., and Verboven, P. (1995). "Precision Cutting and Drilling with a Laser Nd: YAG Slab Laser," *ISEM-11 Conf. Proc.*, pp. 777–784.

Shichun, D., et al. (1995). "A New Technology for Machining Spheres by Rotary, EDM," *ISEM-11*, pp. 921–827.

Simao, J., El-Menshawy, F., Aspinwal, D., Wise, M., and Kockoll, L. (1994). "Simulated Laser Texturing Using a Modified Electrical Discharge Arrangement," *Proc. Seventh Int. Conf. of Production/Precision Engineering (ICPE) and Fourth Int. Conf. on High Technology (JSPE)* Chiba, Japan, pp. 558–563.

Snoeys, R., Staelens, F., and Dekeyser, W. (1986). "Current Trends in Non-conventional Material Removal Processes," *Annals of CIRP*, 35(2):467–480.

Spur, G., Appel, S., and Liebelt, S. (1997). "Non-linear Modelling and Simulation of Laser Cutting and Grooving of Fiber Reinforced Thermo Plastics." *32nd MATADOR Conf.*, Manchester, pp. 381–386.

Stavitskaya, N. B., and Stravinsky, B. L. (1979). "Fundamental Features of Precision Spark Machining," *Electr. Obrab. Mater.*, 4:5–15.

Steigerwald, K., and Meyer, E. (1967). "New Developments in Electron Beam Machining Methods." *Electrical Methods of Machining and Forming. IEE*, Conf. Publ. No. 38, pp. 252–258.

Tabata, N., Yagi, S., and Hishu, M. (1996). "Present and Future of Lasers for Fine Cutting of Metal Plate," *Journal of Material Processing Technology*, 62:309–314.

Tam, S. C., Williams, R., Yang, L. J., Jana, S., Lim, L. E., and Lau, M. W. (1990). "Laser Processing of Air Craft Components," *Journal of Materials Processing Technology*, 32:177–194.

Tani, T., Fukuzawa, Y., and Mohri, N. (2004). "Machining Phenomena in WEDM of Insulating Ceramics," ISEM-XIV *On-site Conf. Proc.* Edinburgh, U.K. (*Journal of Materials Processing Technology*).

Todd, J. A., and Copley, S. M. (1997). "Development of a Prototype Laser Processing System for Shaping Advanced Ceramic Materials." ASME, *Journal of Manufacturing Science and Engineering*, 119:55–67.

Tonshof, H. K., and Dickmann, K. (1989). "Influence of Surface/Subsurface Damage on Magnetic Properties of Laser Cut Electrical Steel Sheets," *Annals of CIRP*, 38(2):695–699.

Tonshof, H. K., and Sturmer, M. (1995). "Machining Using High Energy Radiation," *ISEM-XI Conf.* Switzerland, pp. 49–60.

Tonshof, H. K., Hesse, D., Mommsen, J. (1993). "Micro Machining Using Excimer Laser," *Annals of CIRP*, 42(1):247–251.

Tonshof, K., Egger, R., and Kloclco, F. (1996). "Environmental and Safety Aspects of Electrophysical and Electrochemical Processes," *Annals of CIRP*, 45(2):553–567.

Ueda, T., Yamada, K., and Nakayama, K. (1997). "Temperature of Workpiece Material Irradiated with CO_2 Laser," *Annals of CIRP*, 46(1):117–122.

Yamazaki, M., Suzuki, T., Mori, N., and Kunieda, M. (2004). "EDMing of Microrods by Self Drilled Holes," *ISEM-XIV On-site Conf. Proc.*, Edinburgh, U.K. (*Journal of Materials Processing Technology*).

Yeo, C., Tam, S., Jana, S., and Lau, M. (1994). "A Technical Review of Laser Drilling of Aerospace Materials," *Journal of Materials Processing Technology*, 42:15–49.

Yilbas, B. S. (1996). "Experimental Investigation Into CO_2 Laser Cutting Parameters," *Journal of Materials Processing Technology*, 58:323–330.

www.advancedplasma.com/

www.blkbox.com/~chase/polymer/

www, skion.com/html/technology.htm

www.unl.edu/nmrc/outline.htm

Zhixin, J. (1995). "Study on Mechanical Pulse EDM," *Precision Engineering Journal*, 117(2):89–93.

Chapter 6

Hybrid Electrochemical Processes

6.1 Introduction

In the hybrid electrochemical machining processes, the major material removal mechanism is either chemical dissolution (CD) or electrochemical dissolution (ECD). These machining processes are enhanced by using mechanical machining action or thermal assistance. The combination of these phases leads to high material removal rates and improved surface quality.

In the case of the thermally assisted processes using a laser beam (LB), the local heating of the interelectrode gap enhances the dissolution process in laser-assisted chemical etching (LAE) or laser-assisted electrochemical machining (ECML) during which the current density rises and the dissolution phase becomes more intensive leading to a more productive machining process.

The introduction of a mechanical machining phase assists electrochemical machining (ECM) removal rates by changing the interelectrode gap conditions for the enhanced dissolution process. Under such conditions mechanical depassivation of the machined surface, by removing thin layers of oxides and other compounds from the anode, is ensured. This makes the surface dissolution and hence the smoothening process more intensive (*www.unl.edu/nmrc/hybrid/*).

As can be seen in Fig. 6.1, mechanical abrasion (MA) is combined with ECM to form many hybrid processes such as electrochemical grinding (ECG), electrochemical honing (ECH), and electrochemical superfinishing (ECS).

Furthermore, ultrasonic-assisted electrochemical machining (USMEC) employs an ultrasonic (US) machining component with ECD to improve

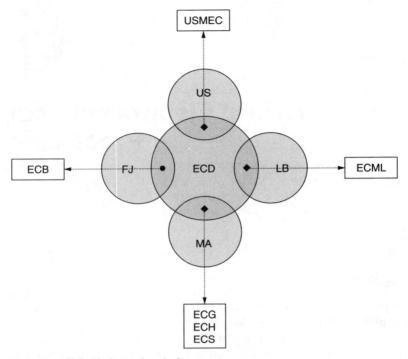

Figure 6.1 Hybrid electrochemical processes.

electrolyte flushing and hence the material removal rate. The mechanical action of the fluid jet (FJ) enhances the CD during electrochemical buffing (ECB).

6.2 Electrochemical Grinding

6.2.1 Introduction

Electrochemical grinding (ECG) utilizes a negatively charged abrasive grinding wheel, electrolyte solution, and a positively charged workpiece, as shown in Fig. 6.2. The process is, therefore, similar to ECM except that the cathode is a specially constructed grinding wheel instead of a cathodic shaped tool like the contour to be machined by ECM. The insulating abrasive material (diamond or aluminum oxide) of the grinding wheel is set in a conductive bonding material. In ECG, the nonconducting abrasive particles act as a spacer between the wheel conductive bond and the anodic workpiece. Depending on the grain size of these particles, a constant interelectrode gap (0.025 mm or less) through which the electrolyte is flushed can be maintained.

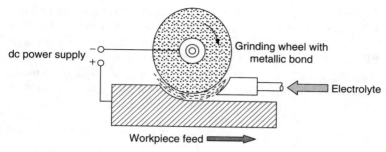

Figure 6.2 Surface ECG.

The abrasives continuously remove the machining products from the working area. In the machining system shown in Fig. 6.3, the wheel is a rotating cathodic tool with abrasive particles (60–320 grit number) on its periphery. Electrolyte flow, usually $NaNO_3$, is provided for ECD. The wheel rotates at a surface speed of 20 to 35 m/s, while current ratings are from 50 to 300 A.

6.2.2 Material removal rate

When a gap voltage of 4 to 40 V is applied between the cathodic grinding wheel and the anodic workpiece, a current density of about 120 to 240 A/cm^2 is created. The current density depends on the material being machined, the gap width, and the applied voltage. Material is mainly removed by ECD, while the MA of the abrasive grits accounts for an additional 5 to 10 percent of the total material removal.

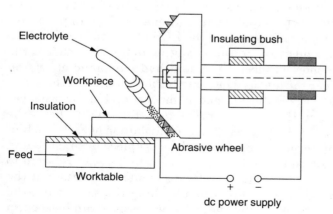

Figure 6.3 ECG machining system components.

Removal rates by ECG are 4 times faster than by conventional grinding, and ECG always produces burr-free parts that are unstressed. The volumetric removal rate (VRR) is typically 1600 mm³/min. McGeough (1988) and Brown (1998) claimed that to obtain the maximum removal rate, the grinding area should be as large as possible to draw greater machining current, which affects the ECD phase. The volumetric removal rate (mm³/min) in ECG can be calculated using the following equation:

$$\text{VRR} = \frac{\varepsilon I}{\rho F}$$

where ε = equivalent weight, g
I = machining current, A
ρ = density of workpiece material, g/mm³
F = Faraday's constant, C

The speed of penetration of the grinding wheel into the workpiece, V_g (mm³/min), is given by Kalpakjian (1997) as follows:

$$V_g = \frac{\varepsilon}{\rho F} \frac{v}{g_w K_p} \kappa$$

where v = gap voltage, V
g_w = wheel-workpiece gap, mm
K_p = coefficient of loss (1.5–3)
κ = electrolyte conductivity, $\Omega^{-1}\text{mm}^{-1}$

ECG is a hybrid machining process that combines MA and ECD. The machining rate, therefore, increases many times; surface layer properties are improved, while tool wear and energy consumption are reduced. While Faraday's laws govern the ECD phase, the action of the abrasive grains depends on conditions existing in the gap, such as the electric field, transport of electrolyte, and hydrodynamic effects on boundary layers near the anode. The contribution of either of these two machining phases in the material removal process and in surface layer formation depends on the process parameters. Figure 6.4 shows the basic components of the ECG process. The contribution of each machining phase to the material removal from the workpiece has resulted in a considerable increase in the total removal rate Q_{ECG}, in relation to the sum of the removal rate of the electrochemical process and the grinding processes Q_{ECD} and Q_{MA}, when keeping the same values of respective parameters as during the ECG process. As can be seen in

Hybrid Electrochemical Processes   185

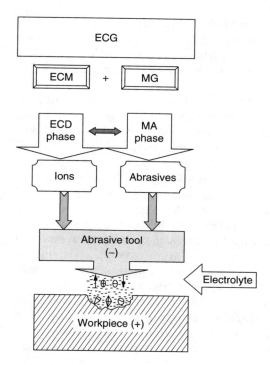

Figure 6.4 ECG process components.

Fig. 6.5, the introduction of MA, by a rotary conductive abrasive wheel, enhances the ECD process. The work of the abrasive grains performs the mechanical depolarization by abrading the possible insoluble films from the anodic workpiece surface. Such films are especially formed in case of alloys of many metals and cemented carbides. A specific purpose of the abrasive grains is, therefore, to depassivate mechanically the workpiece surface. In the machining zone there is an area of simultaneous

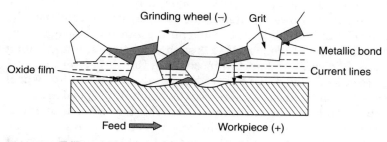

Figure 6.5 ECD and MA in the machining gap during ECG.

ECD and MA of the workpiece surface, where the gap width is less than the height of the grain part projecting over the binder. Another area of pure electrochemical removal where the abrasive grains do not touch the workpiece surface exists at the entry and exit sides of the wheel.

The increase in performance of the ECG process results from interactions between MA, ECD, and changes in surface layer properties of material in the machining area. The increase in longitudinal feed rate and wheel speed and the decrease in voltage result in a smaller gap width and, consequently, in an increase in the MA contribution to the metal removal process. A decrease in the feed rate is followed by a considerable reduction in cutting forces, down to the values near zero at the transition to "pure" ECD. Machining conditions at which the MA disappears depend on the electrical parameters, electrochemical machinability of the material in a given electrolyte, and grinding wheel features, especially grain size and height. Generally slow feed rates produce larger overcut, poor surface finish, and wider tolerances, while excessive wheel wear occurs as a result of a feed rate that is too fast. Use of expensive diamond grinding wheels considerably reduces the grinding forces which in turn significantly reduces the abrasive wear and provides notable savings in machining costs. The MA phase increases the process productivity, dimensional accuracy, and surface integrity while also reducing energy consumption. When machining steels, MA plays a secondary role since it accounts for only 15 percent of the removal rate. In case of sintered carbides, that contribution reaches 30 percent (Kaczmarek, 1976).

For the electrolytic dissolution part, ECM variables such as material type, electrolyte, and current density determine the rate of material removal. For high machining rates, the greatest current density is to be employed. However, such current density is limited by the anodic dissolution rate for a particular alloy-electrolyte system and the boiling point of the electrolyte (*Machining Data Handbook*, 1997). The interaction between the abrasion action and electrolytic dissolution affects the removal rate of the machined material. By increasing the wheel unit pressure while removing a thicker layer, it reduces the size of the machining gap that allows for higher current density. In contrast, the larger the grain size, the wider is the machining gap and, consequently, the lower is the current density. Under such conditions the machining rate is impaired despite the partial increase in the abrasion rate.

Rumyantsev and Davydov (1984) showed also an optimum wheel speed where the removal rate and hence the machining speed are maximum. Figure 6.6 shows the maximum attainable feed rate, as an index of machinability. Accordingly, aluminum and copper alloys have the highest machinability, while stainless steel and carbides have the lowest one (El-Hofy, 1995).

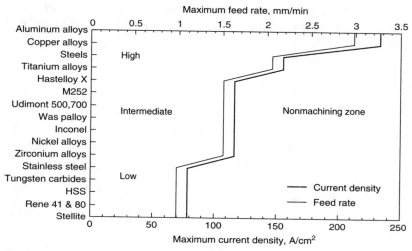

Figure 6.6 Machinability ratings in ECG (El-Hofy, 1995).

6.2.3 Accuracy and surface quality

Traditional grinding removes metal by abrasion, leaving tolerances of about ±0.003 mm and creating heat and stresses that make grinding thin stock very difficult. In ECG however a production tolerance of ±0.025 mm is easily obtainable. Under special circumstances a tolerance of ±0.008 mm can be achieved. The ability to hold closer tolerances depends upon the current, electrolyte flow, feed rate, and metallurgy of the workpiece itself. Accuracies achieved are usually about ±0.125 mm. A final cut is usually done mostly by the grinding action to produce a good surface finish and closer dimensional tolerances. It is recommended that lower voltages be used for closer tolerances, reduced overcut, sharp edges, and a bright surface finish. ECG can grind thin material of 1.02 mm, which normally warp by the heat and pressure of the conventional grinding thus making closer tolerances difficult to achieve. In ECG there is little contact between the wheel and workpiece, which eliminates the tendency of the workpiece to warp as it might with orthodox grinding (Brown, 1998).

The main drawback of ECG is the loss of accuracy when the inside corners are ground. Because of the electric field effect, radii better than 0.25 to 0.375 mm can seldom be achieved. The reason for this problem is that the point of highest pressure of the electrolyte is the wheel corner. However, high-speed grinding benefits both inside and outside corners. The surface finish produced varies from 0.2 to 0.3 µm depending on the metal being machined, abrasive grit size and wheel speed and feed rate. For better surface quality a finish pass at a low voltage of 3 to 5 V and relatively high speed (250–500 mm/min) is recommended (Shpitalni, 1981).

6.2.4 Applications

The ECG process is particularly effective for

1. Machining parts made from difficult-to-cut materials, such as sintered carbides, creep-resisting (Inconel, Nimonic) alloys, titanium alloys, and metallic composites.
2. Applications similar to milling, grinding, cutting off, sawing, and tool and cutter sharpening.
3. Production of tungsten carbide cutting tools, fragile parts, and thin-walled tubes.
4. Removal of fatigue cracks from steel structures under seawater. In such an application holes about 25 mm in diameter, in steel 12 to 25 mm thick, have been produced by ECG at the ends of fatigue cracks to stop further development of the cracks and to enable the removal of specimens for metallurgical inspection.
5. Producing specimens for metal fatigue and tensile tests.
6. Machining of carbides and a variety of high-strength alloys.

The process is not adapted to cavity sinking, and therefore it is unsuitable for the die-making industry.

6.2.5 Advantages and disadvantages

Advantages

- Absence of work hardening
- Elimination of grinding burrs
- Absence of distortion of thin fragile or thermosensitive parts
- Good surface quality
- Production of narrow tolerances
- Longer grinding wheel life

Disadvantages

- Higher capital cost than conventional machines
- Process limited to electrically conductive materials

- Corrosive nature of electrolyte
- Requires disposal and filtering of electrolyte

6.3 Electrochemical Honing

6.3.1 Introduction

Electrochemical honing (ECH) combines the high removal characteristics of ECD and MA of conventional honing. The process has much higher removal rates than either conventional honing or internal cylindrical grinding. In ECH the cathodic tool is similar to the conventional honing tool, with several rows of small holes to enable the electrolyte to be introduced directly to the interelectrode gap. The electrolyte provides electrons through the ionization process, acts as a coolant, and flushes away chips that are sheared off by MA and metal sludge that results from ECD action. The majority of material is removed by the ECD phase, while the abrading stones remove enough metal to generate a round, straight, geometrically true cylinder. During machining, the MA removes the surface oxides that are formed on the work surface by the dissolution process. The removal of such oxides enhances further the ECD phase as it presents a fresh surface for further electrolytic dissolution. Sodium nitrate solution (240 g/L) is used instead of the more corrosive sodium chloride (120 g/L) or acid electrolytes. An electrolyte temperature of 38°C, pressure of 1000 kPa, and flow rate of 95 L/min can be used. ECH employs dc current at a gap voltage of 6 to 30 V, which ensures a current density of 465 A/cm^2 [Randlett et al. (1968)]. Improper electrolyte distribution in the machining gap may lead to geometrical errors in the produced bore.

6.3.2 Process characteristics

The machining system shown in Fig. 6.7 employs a reciprocating abrasive stone (with metallic bond) carried on a spindle, which is made cathodic and separated from the workpiece by a rapidly flowing electrolyte. In such an arrangement, the abrasive stones are used to maintain the gap size of 0.076 to 0.250 mm and, moreover, depassivate the machining surface due to the ECD phase occurring through the bond. A different tooling system (Fig. 6.8) can be used where the cathodic tool carries nonconductive honing sticks that are responsible for the MA. The machine spindle that rotates and reciprocates is responsible for the ECD process.

The material removal rate for ECH is 3 to 5 times faster than that of conventional honing and 4 times faster than that of internal cylindrical grinding. Tolerances in the range of ±0.003 mm are achievable,

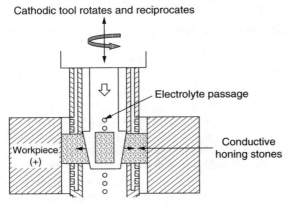

Figure 6.7 ECH schematic.

while surface roughnesses in the range of 0.2 to 0.8 μm R_a are possible. To control the surface roughness, MA is allowed to continue for a few seconds after the current has been turned off. Such a method leaves a light compressive residual stress in the surface. The surface finish generated by the ECH process is the conventional cross-hatched cut surface that is accepted and used for sealing and load-bearing surfaces. However, for stress-free surfaces and geometrically accurate bores, the last few seconds of MA action should be allowed for the pure ECD process.

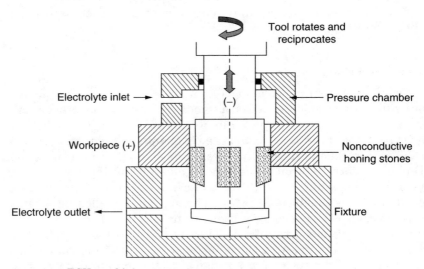

Figure 6.8 ECH machining system components.

6.3.3 Applications

As a result of the rotating and reciprocating honing motions, the process markedly reduces the errors in roundness through the rotary motion. Moreover, through tool reciprocation both taper and waviness errors are also reduced as shown in Fig. 6.9. Because of the light stone pressure used, heat distortion is avoided. The presence of the ECD phase introduces no stresses and automatically deburrs the part. ECH can be used for hard and conductive materials that are susceptible to heat and distortion. The process can tackle pinion gears of high-alloy steel as well as holes in cast tool steel components. Hone forming (HF)

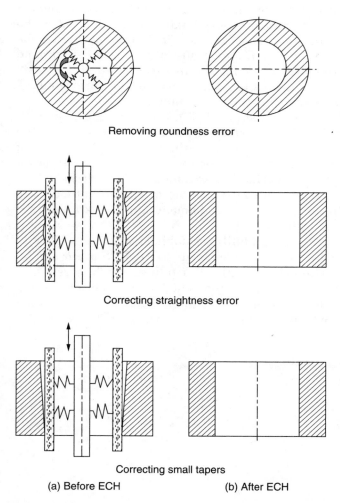

Figure 6.9 ECH effects on bore errors.

is an application that combines the honing and electrodeposition processes. It is used to simultaneously abrade the work surface and deposit metal. In some of its basic principles the method is the reversal of ECH. According to the *Metals Handbook* (1989), this method is used in case of salvaging parts that became out-of-tolerance and reconditioning worn surfaces by metal deposition and abrasion of the new deposited layers.

6.4 Electrochemical Superfinishing

6.4.1 Introduction

Conventional superfinishing by vibration grinding is a microfinishing operation in which the surface mirco-irregularities are removed by the continuous and slow reciprocation of abrasive sticks that move along the workpiece length. The sticks oscillate concurrently with short and rapid strokes with a continuously revolving workpiece as shown in Fig. 6.10. This process is however known to sustain some of the surface microirregularities such as waviness and out of roundness.

In electrochemical superfinishing (ECS), the combination of electrolytic dissolution (ECD) and mechanical scrubbing (MS) improves the performance of the conventional superfinishing process. As a result of such a combination, the dissolution process assists the small stock removal rate due to the mechanical chipping action. Higher stock removal rates become achievable using either a separate cathodic tool electrode, as shown in Fig. 6.11, or a diamond abrasive stick with a metallic bond, as shown in Fig. 6.12.

The high stock removal capabilities combined with the ability to generate close dimensions gave high merits to the ECS process in all fields of industry. The need for initial grinding, which is required before conventional superfinishing, is avoided. ECS can be used when other processes fail to yield high removal rates or generate the required size in difficult-to-machine alloys as well as tool steel. Applying ECS

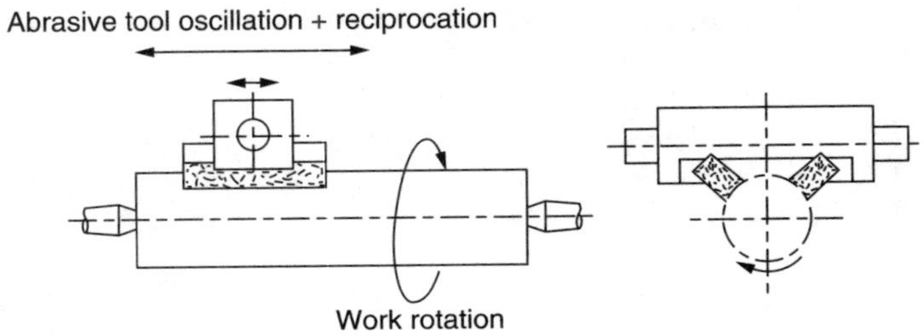

Figure 6.10 Vibratory superfinishing schematic.

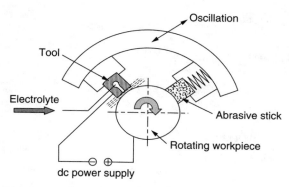

Figure 6.11 Superfinishing using separate EC electrode (Youssef, 1971).

to parts that are susceptible to heat and distortion is advantageous because the bulk of the metal is removed electrochemically in an electrolyte-cooled atmosphere. The problem of thermal distortion, normally found in conventional superfinishing, is therefore eliminated. Burr-free components can also be obtained as a result of the ECD process.

6.4.2 Material removal process

In ECS the dissolution action is accompanied by the formation of a protective oxide film on the anodic surface. The MA scrubs away any high spots protruding from the ideal configuration. These spots, with fresh metal contacting the electrolyte, will be subjected to a heavier ECD phase compared to areas still covered with the protective film as shown

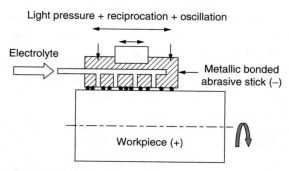

Figure 6.12 ECS using metallic bond diamond stick (Youssef, 1971).

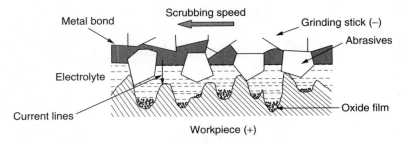

Figure 6.13 Metal removal mechanisms in ECG.

in Fig. 6.13. Under such circumstances, the protecting film can be used to correct the geometric inaccuracies such as cylindericity and roundness errors.

The power of the oxide film depends on the electrolyte used. Some electrolytes possess fairly strong power to reduce the ECD with their protective film. Other electrolytes have a strong protective film too, but the electric charge needed to build up the film is too small. According to El-Lawendy (1977), inspection of the effect of ECD without mechanical scrubbing showed the presence of a dark, visible film that reduced the current by 10 to 20 percent and the metal removal rate by 50 percent. In some cases, mechanical scrubbing should, therefore, alternate with the ECD process.

In order to avoid metallurgical damage that may occur by mechanical scrubbing, light stone pressure is used after ECM to produce a bright surface finish, tolerances of about ±0.013 mm on the diameter, and a roundness and straightness of less than 0.007 mm. It has been reported by Datta and Landolt (1983) that the use of pulsating voltage allows for the application of high instantaneous current densities. This is possible because each current pulse is followed by a relaxation time of zero current, which allows for the removal of reaction products and the heat generated by the Joule effect from the interelectrode gap.

The process variables include all those related to ECM parameters, such as gap voltage, electrolyte type, concentration, and temperature. Regarding the MA, stone pressure, oscillation frequency, and amplitude besides the abrasive grain characteristics markedly affect the process performance. Experiments conducted by El-Hofy (1990) using pulsating voltage showed that the linear removal rate increases with current density as shown in Fig. 6.14. For the given scrubbing speed, the MS remains unchanged. Since both the dissolution phase and MS action adds to the total removal rate, it is clear that a higher ECD contribution can be obtained at higher current densities. Under such conditions, more energy becomes available for the dissolution phase to prevail. A rise in the scrubbing speed, voltage, and duty cycle would lead to an increase in the

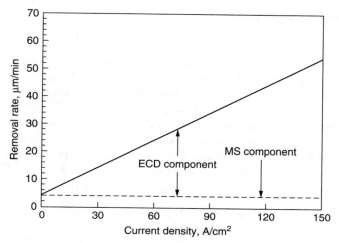

Figure 6.14 ECS removal rates at different current densities (El-Hofy, 1990).

removal rate. The high energy available enhances the oxide film removal process with a consequent rise in the dissolution phase. The percentage contribution of the dissolution phase varies between 0 at 20 percent duty cycle to about 95 percent when using a 100 percent duty cycle. As the scrubbing speed increases, the percentage of the MA phase increases.

6.4.3 Process accuracy

Regarding the workpiece out-of-roundness, Fig. 6.15 shows a reduction of roundness error from 24 to 8 µm during 2 min of machining using 19 V, a 67 percent duty cycle, and a scrubbing speed of 18.55 m/min. Similarly, surface roughness has been decreased from 2.25 to 0.65 µm R_a (El-Hofy, 1990).

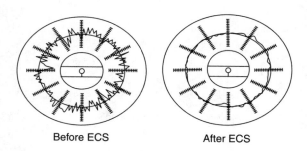

Figure 6.15 Effect of ECS on out-of-roundness profiles.

6.5 Electrochemical Buffing

6.5.1 Introduction

Mechanical buffing is a slow finishing process used for achieving smooth, bright, and mirrorlike surfaces. The process is carried out under dry conditions, which raises dust and makes the working environmental conditions unsuitable. Electrochemical buffing (ECB), shown in Fig. 6.16, uses a carbon fiber cloth that rubs the anodic specimen against a revolving cathode fiber buff. Electrolytes of NaCl or $NaNO_3$ are supplied to the machining zone using a suitable pump. The machining current flows from the workpiece to the cathode through the carbon cloth. Typical ECB conditions have been reported by Hoshino et al. (1995):

Electrolyte type	NaCl 150 g/L + $NaNO_3$ 10 g/L
	NaCl, 100–500 g/L
Current density	0.1–7 A/cm^2
Buffing speed	1–6 m/s
Contact pressure	100–600 g
Electrolyte flow rate	1–10 L/min

6.5.2 Material removal process

ECD of the anodic specimen mainly takes place on the surface of the specimen where it is rubbed by the carbon cloth buff. The current density, the type of electrolyte, and the workpiece material control the polishing speed. For high-speed polishing, an NaCl electrolyte is used where high current density is ensured. The addition of Al_2O_3 abrasives

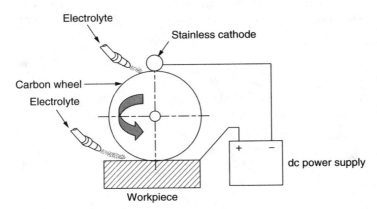

Figure 6.16 ECB schematic.

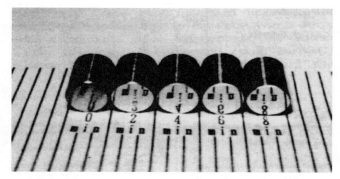

Figure 6.17 Specimens polished by ECB (Hoshino et al. 1995).

(200 mesh number) to the machining medium increases the amount of material removal; however, surface smoothing and brightness are decreased. During ECB, a passive oxide film is normally formed on the surface of the stainless steel workpiece. MA and hence the removal of such a film on the high spots of the surface irregularities enhances the dissolution phase to prevail in these spots, which in turn makes the surface smoother and brighter (Fig. 6.17).

6.6 Ultrasonic-Assisted ECM

6.6.1 Introduction

The demand for machining hard and brittle materials is steadily increasing in many applications. Ultrasonic machining (USM) produces parts having better surface quality. However, the material removal rate and hence the machining productivity is low. On the other hand ECM has the advantage of achieving high machining rates as well as better surface quality. ECM is only effective for machining conductive materials, while USM is suitable for hard and brittle materials such as ceramics. However, these two processes are not effective when machining composite materials that contain a mixture of metallic and nonmetallic phases.

Merging the two processes has the advantage of combining their virtues especially when tackling difficult-to-machine composite materials. Ultrasonic-assisted electrochemical machining (USMEC) combines both ECM for removing the metallic conducting parts and USM for removing the nonconducting hard and brittle phases. In this regard, Clifton et al. (1993) employed USMEC to overcome the obstructions posed to ECM, by the presence of paint, marine fouling, and slag inclusions.

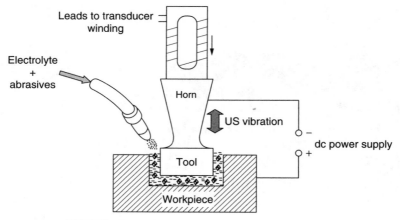

Figure 6.18 USMEC schematic.

The machining system for this hybrid process, shown in Fig. 6.18, employs a normal USM. Electrolyte replaces water as an abrasive carrier liquid. A voltage of 3 to 15 V dc is normally used and ensures current densities between 5 and 30 A/cm^2. Besides the dissolution process, the machine head and hence the cathodic tool are vibrated at the ultrasonic frequency of 20 kHz and an amplitude of 8 to 30 μm. USMEC process parameters include those related to ECM and USM. Figure 6.19 shows the main components of the process.

6.6.2 Material removal process

During USMEC, the intensity of the dissolution phase depends on the relative position of the tool with respect to the workpiece. This phase reaches its maximum level at a gap size equal to the size of the statically pressed abrasive grains. Under this particular condition, the nonconductive abrasive grains form the minimum interelectrode gap size. When a pulsed voltage replaces the straight dc one, it must be synchronized with the tool oscillation in order to maintain an efficient ECD and avoid the formation of spark discharges across the interelectrode gap.

The dissolution phase occurs along with the MA caused by the ultrasonic impact of abrasive grains at the machined surface. Since the anodic dissolution phase is accompanied by the formation of a brittle (passive) oxide layer that hinders further dissolution, the abrasive grains are, therefore, expected to act not on the material of the workpiece but mainly on the brittle oxide layer itself. Under such circumstances, the dissolution process is enhanced due to the mechanical depassivation of the machined surface. The efficiency of both processes is, therefore, improved

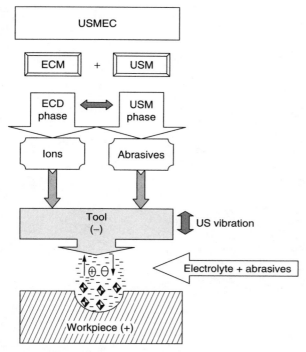

Figure 6.19 USMEC components.

in terms of higher machining speeds and lower tool wear, compared to normal USM. The presence of the dissolution phase conditions the chipping marks formed by the MA action, and hence the improvement of surface quality is also expected.

The increase in current density enhances the machining productivity and reduces the tool wear. The produced accuracy is reduced possibly due to the side-machining effect, which leads to larger side-machining gaps than those expected in the case of pure USM. Under such circumstances the side tool wear is greatly reduced.

6.7 Laser-Assisted ECM

ECM is a dissolution process, which is carried out mostly at temperatures of 20 to 80°C. During machining only dissolution action occurs in the machined surface. It is possible to increase the rate of ECM by increasing the temperature of the interelectrode gap. In this regard, the application of a laser beam to the machined surface is an efficient way for increasing the temperature of the machining zone during laser-assisted electrochemical machining (ECML). The wavelength of laser radiation

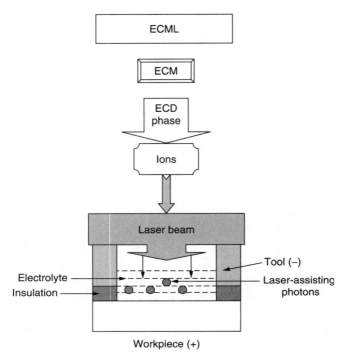

Figure 6.20 ECML schematic.

should be chosen in such a way as to minimize the energy adsorbed by the electrolyte layer, which should be as thin as possible. The heat of the laser beam causes many physical and chemical phenomena on the machined surface as well as on the surface layer of the material (Fig. 6.20).

The laser radiation affects the rate of electrochemical reactions by several mechanisms. Local heating of the workpiece area subjected to the laser beam leads to changes in the equilibrium or steady-state potential (open circuit potential). It increases the rate of the reaction that proceeds under the charge-transfer control, accelerates the mass transfer, changes the current efficiency, decreases the metal passivity degree, and finally increases the concentration of current carriers during ECM of semiconductors. The effect of laser radiation on the current efficiency of the anodic dissolution is not well understood. If the current efficiency increases with temperature, an increase of the degree of process localization will be especially pronounced. In contrast, a marked decrease of the current efficiency with an increase in temperature may present an impenetrable barrier to the use of laser in some cases of ECM. The formation of a high-adsorptive layer of dissolution products may be a barrier to the use of ECML.

The experimental work of Zybura-Skrabalak and Ruszaj (2000) showed that with a temperature increase from 20 to 40°C, the current density and the intensity of dissolution increase from 0.6 to 1.8 mm^3/(min·A). Laser assistance of ECM can be effectively used for improvement in the micromachining processes by increasing the material removal rate of normal ECM by 10 times. It causes localization of the anodic dissolution, especially when the passivation process occurs during electrochemical dissolution. One way to further increase the localization of laser action is to use pulse radiation. An increase of the electrolyte flow rate has a similar effect. The prospective application of ECML can be useful in the shaping of small elements (5–500 µm) with high accuracy (1–10 µm) for the electronic and space industries especially when the elements are made from difficult-to-machine materials such as alloys, composites, and ceramics. Pajak and coworkers (2004) reported that laser-assisted jet ECM (LAJECM) improves the machining efficiency and precision of small holes compared to that of single-jet ECM.

References

Brown, J. (1998). *Advanced Machining Technology Handbook*. New York: McGraw-Hill.
Clifton, D., Imai, Y., and McGeough, J. (1993). "Some Ultrasonic Effects in Machining Materials Encountered in the Offshore Industries," *30th MTDR Conf.*, pp. 119–123.
Datta, M., and Landolt, D. (1983). "Electrochemical Saw Using Pulsating Voltage," *Journal of Applied Electrochemistry*, 13:795–801.
El-Hofy, H. (1990). "Characteristics of Pulsed EC-Superfinishing," *Alexandria Engineering Journal (AEJ)*, 29(1):83–100.
El-Hofy, H. (1995). "Machinability Indices for Some Non Conventional Machining Processes," *Alexandria Engineering Journal (AEJ)*, 34(3):231–245.
El-Lawendy, M. (1977). "Electrochemical Super Finishing," M.Sc. Thesis, Alexandria University.
Hoshino, S., and Ogawa, K. (1995). "Electrochemical Buffing Using Carbon Fiber Cloth," *ISEM 11*, pp. 577–583.
Kaczmarek, J. (1976). *Principles of Machining by Cutting, Abrasion and Erosion*, Stevanage, U.K.: Peter Peregrines.
Kalpakjian, S. (1997). *Manufacturing Processes for Engineering Materials*, Menlo Park, CA: Addison Wesley.
Machining Data Handbook. (1997). 3rd ed. Cincinnati, OH: Machinability Data Center, Institute of Advanced Manufacturing.
McGeough, J. A. (1988). *Advanced Methods of Machining*. London. New York: Chapman and Hall.
Metals Handbook. (1989). Vol. 16, *Machining*. Materials Park, OH: ASM International.
Pajak, P. T., DeSilva, A., McGeough, J., and Harrison, D. (2004). "Modeling the Aspects of Precision and Efficiency in Laser Assisted Jet Electrochemical Machining (LAJECM)," *ISEM XIV, On-site Conf. Proc.*, Edinburgh, U.K. (*Journal of Materials Processing Technology*).
Randlett, E., et al. (1968). "Electrochemical Honing (ECH)," Technical Paper MR68–815, 9 pp.
Rumyantsev, E., and Davydov, A. (1984). *Electrochemical Machining of Metals*. Moscow: MIR Publishers.
Shpitalni, M., et al. (1981). "Adaptive Control System for Optimising the ECG Process under the Over Cut Constraint," *Annals of CIRP*, 30(1):97–102.
Youssef, H. A. (1971). "Electrolytic Superfinishing of Sintered Carbides," Faculty of Engineering Bulletin, Alexandria University, X:209–231.
Zybura-Skrabalak, M., and Ruszaj, A. (2000). "Investigation Aiming to Increase the Rate of Electrochemical Dissolution Process," *CAPE-2000 Conf.*, Edinburgh, U.K., pp. 163–174.
www.unl.edu/nmrc/hybrid/hybrid_machining.htm

Chapter 7

Hybrid Thermal Processes

7.1 Introduction

In hybrid thermal machining; the major material removal mechanism is a thermal one which normally leads to melting and evaporation of the workpiece material. Thermal machining can be assisted using electrochemical dissolution (ECD) and/or mechanical abrasion (MA). This combination leads to high removal rates and improved product quality.

In electrochemically assisted processes, the combination of the electrodischarge erosion (EDE) phase with the ECD phase forms the new electrochemical discharge machining (ECDM), [also known as electroerosion dissolution machining (EEDM) or electrochemical arc machining (ECAM)] process that occurs in electrolyte solutions using pulsed voltage. Here the dissolution phase enhances the discharging process which adds to the material removal.

Mechanically assisted processes combine the mechanical abrasion action with the thermal one in case of electrodischarge grinding (EDG) and abrasive electrodischarge machining (AEDM), which employ solid or loose abrasives in dielectric liquids. Electrodischarge machining with ultrasonic assistance (EDMUS) introduces ultrasonic vibration to the tool that enhances the electrodischarging action. The cavitation generated by such a vibration enhances EDM by improving the dielectric flushing, and hence the material removal from the surface craters, during microdrilling and production of slots and grooves.

A triplex hybrid machining adds to the material removal and is also achievable by combining the EDE phase to the ECD and MA processes in electrochemical discharge grinding (ECDG). Similarly, electrochemical dissolution and mechanical friction provide reasonable assistance to the discharging process during brush erosion-dissolution mechanical

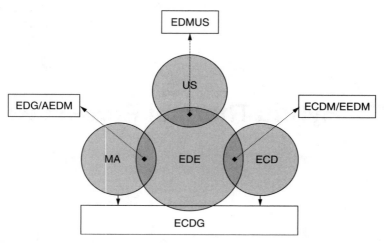

Figure 7.1 Hybrid thermal machining processes.

machining (BEDMM). Figure 7.1 shows the various hybrid thermal machining processes.

7.2 Electroerosion Dissolution Machining

Novel methods of machining hard metals, which are difficult to cut by conventional methods, continue to attract attention. Electrochemical machining and electrodischarge machining have proven to be very useful. However, drawbacks such as the expense of tooling for machining large cavities, the high cost of machining systems, low rates of metal removal, and the presence of a recast layer, which often has to be removed in EDM, have hindered wider acceptance of these techniques.

EEDM (also called ECDM or ECAM) is a new development, which combines features of both ECD and EDE. It utilizes electrical discharges in electrolytes for material removal. Such a combination allows high metal removal rates to be achieved. EEDM has found a wide range of applications in the field of wire cutting, hole drilling, and finishing of dies and molds by El-Hofy and McGeough (1988), Crichton and McGeough (1984), and Levy et al. (1983). Further applications regarding machining of composites using NaOH electrolyte have been reported by Jain et al. (1991).

The EEDM process is a further development of pulsed electrochemical machining (PECM) where, according to Saushkin, et al. (1982), at high input power, phenomena that limit further dissolution may arise. Under such circumstances, the machining medium changes to a gas-vapor mixture that interferes with the ion transfer in the electric field.

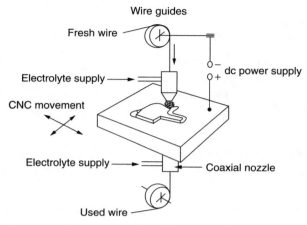

Figure 7.2 Electroerosion dissolution wire machining.

If the field strength is high enough to cause gap breakdown, the nature of charge transfer is altered causing the combined action of EEDM. Such a machining process is associated with a glow appearing in the interelectrode gap. The glow and the subsequent breakdown of the interelectrode gaps were localized at points where the gas content and temperature of the working medium are likely to be highest.

The machining system for EEDM wire cutting is shown in Fig. 7.2. It adopts pulsed voltage and liquid electrolytes as the machining medium that ensures the occurrence of ECD along with the discharge phase (El-Hofy, 1995). A further arrangement used in hole drilling by Khayry and El-Hofy (1990) is shown in Fig. 7.3. A full wave rectified voltage is applied during the vibration of the workpiece or the tool at 100 Hz and an amplitude and phase shift with respect to the voltage as shown in Fig. 7.4. The tool feed, vibration amplitude, and phase

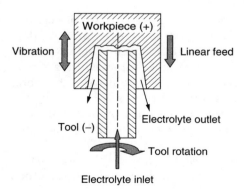

Figure 7.3 EEDM drilling.

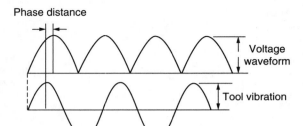

Figure 7.4 Voltage and tool-oscillation waveform (El-Hofy and McGeough, 1988).

angle determine the instantaneous machining gap width and hence the intensity and duration of each phase.

EEDM is affected by many variables that control its performance, accuracy, and surface quality. Among these are the electrical parameters such as pulse time, relaxation interval, and pulse current. Workpiece characteristics such as melting point and specific heat have been dealt with by El-Hofy (1992, 1996a). During EEDM, the machining medium is subjected to varying contamination conditions. This is caused by many interfering phenomena such as gas generation and varying dissolution intensity. The presence of different types of pulses that are responsible for the erosion phase in the form of metal resolidified particles, and the change of electrolyte flow rate, are possible causes of gap contamination. The breakdown characteristics change with time, and hence the initially required dimensional accuracy cannot be reached anymore. The gas-liquid wedge, formed during electrolysis, is the determining factor in this particular hybrid machining process. In this regard, it has been found that the superposition of low-voltage pulse components (Fig. 7.5) stabilizes the machining process and makes it possible to reduce the breakdown voltage, enhance the machining productivity, and reduce the surface roughness.

Figure 7.6 presents the main machining phases and process components of EEDM. According to Fig. 7.7, spark discharges occur at random locations across the machining gap while electrolysis is believed to be localized in the proximity of the pits of the formed craters which are soon made smooth, probably as a result of the high temperature of the metal and electrolyte. The EEDM material removal rate is enhanced by the sparking action and not by the arcing one because the latter usually results in a low and localized material removal rate and yields more irregular machined surfaces.

Since EEDM relies on machining by EDE assisted by ECD, surface properties are expected to be electrodischarge machined and smoothed

Hybrid Thermal Processes

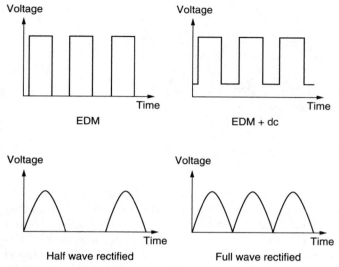

Figure 7.5 Voltage waveforms.

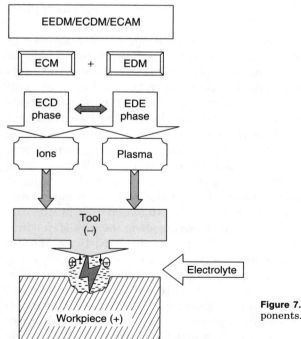

Figure 7.6 EEDM process components.

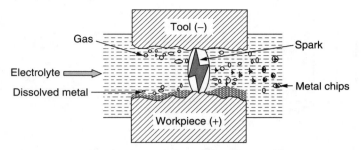

Figure 7.7 Surface formations in EEDM (El-Hofy and Khayry, 1995).

by the ECD action, or electrochemically machined with the surface conditioned by electrodischarge craters (El-Hofy and Khayry, 1995). In this regard Levy et al. (1983) noticed that the depth of the thermally affected layer was comparatively low at low discharge intensity due to the ECD occurring at high current density. Moreover, in hole drilling, the damaged layer was extended to only one-third of the hole depth from the exit side.

The thickness of the heat-affected layer was reported to be 290 µm. Khayry (1984) and Levey et al. (1983) reported the existence of microcracks. Some alloying of the surface layers with the tool material was also observed at high voltages, which may enhance the mechanical properties of the material.

The increase of the machining rate raises the volumetric removal rate (VRR) due to the enhanced erosion process. Such an increase reaches a maximum level and then decreases probably due to the narrow interelectrode gap and the rise in the gap contamination level together with the percentage of gas-vapor mixture. The machining rate increases with the electrolyte flow velocity. At small machining rates, the effect of flow velocity on the VRR is greatly reduced as a result of the wide interelectrode gaps that exist under such machining conditions. Consequent gap short circuiting and the termination of the machining process follow a large increase in machining rate. PECM prevails at lower feed rates that avoid gap breakdown and the occurrence of discharges (McGeough and coworkers, 1983). Between PECM and the shorting limit, the possible EEDM zone covers a wider range of machining rates as the electrolyte flow velocity is raised (El-Hofy, 1996b). PECM occurs at a gap pressure of 0.1 MPa and a current density of 10 A/cm^2. The increase in gap pressure, electrolyte velocity, and current density enhances the erosion process. It should be borne in mind that a rise in the machining voltage allows for more energy to be available for an intensified dissolution process, gas generation, and consequently the discharging action. During EEDM drilling, the specific material removal rate increases with the gap voltage. According to

McGeough and coworkers (1983), the removal rate is much greater than that calculated by pure electrochemical machining (ECM).

The frequency of pulses markedly affects the functional dependence of the gas content, which initiates a discharging phenomenon. There is a pulse frequency below which pulsed ECM occurs. A further increase in pulse frequency, beyond this limit, leads to the erosion process (El-Hofy and McGeough, 1989). During EEDM, sparks, arcs, open circuits, and micro short circuit pulses occur in the electrolytic medium. Accordingly, open circuits and the rise time of arcs and sparks dissolves the metal by ECD and generates enough gases for discharges to take place during the rest of the time of both arcs and sparks. Micro short circuit pulses are nonproductive with respect to ECD or EDE phases. The percentage of these pulses reaches 35 percent and it reflects the status of the gap contamination level, machining accuracy, and surface quality. Micro short circuits are markedly affected by machining conditions such as the feed rate, pulse parameters, as well as the electrolyte flushing conditions (El-Hofy, 1995).

The surface generation process occurs through electrolytic dissolution of varying intensity, depending on the gap size and the consequent crater formation at random locations over the entire machined surface. Micrographs have indicated that electrolysis is apparently localized in the proximity of the pits of the craters that are soon made smooth, probably as a result of the high temperature of the metal and the electrolyte in this zone (Saushkin et al., 1982). It should be mentioned here that due to the presence of ECD action, better surfaces than those provided by EDM are possible.

The general appearance of the machined surface constitutes less turbulence than that reported with EDM. The work of Saushkin et al. (1982) revealed that the gas-liquid layer forms at an average current density of 15 to 20 A/cm^2 and the superposition of the low-voltage pulse component stabilizes the hybrid machining process and makes it possible to reduce the breakdown voltage and enhance the discharging process. The results of El-Hofy (1992) indicated that the surface roughness, produced by EED wire cutting, can be expressed by

$$R_a = 1.19 V_s^{0.111} t_p^{0.0023}$$

where V_s is the feed rate (mm^2/min) and t_p is the pulse duration (µs).

Considering the random nature encountered during the surface generation mechanism, the crater depth, volume, diameter-to-depth ratio, and roughness have been calculated using the surface roughness profiles. It was accordingly evident that with the hybrid EEDM process the roughness indices are markedly reduced and there is an absence of the heat-affected layer (El-Hofy and coworkers, 1990).

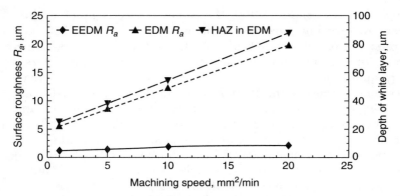

Figure 7.8 Effect of machining speed on the surface roughness and the damaged layer in EDM and EEDM.

The increase in electrolyte flow velocity during EEDM raises the surface roughness. Higher electrolyte flow velocities increase the ignition delay time of pulses. Under such circumstances, more time is allowed for the dissolution process to occur at higher voltages. Consequently, larger quantities of gas are generated and cause severe discharges, larger craters are formed, and rougher surfaces are produced. Kuneida and Furuoya (1991) found that the stock removal rate is increased due to the enlarged volume of discharge craters and the more frequent occurrence of discharges caused by the oxygen gas in the EDM gap.

The increase in feed rate reduces the size of the interelectrode gap, which intensifies the dissolution phase (since more gas evolves and encourages the discharging action) and leaves larger craters on the machined surface. The increase in machining rate would, therefore, lead to a rise in surface roughness as described in Fig. 7.8. The increase of pulse energy ensures the presence of more energy for an efficient dissolution phase through the machining gap, and hence larger rates of removal and rougher surfaces are expected.

The machinability of metals and alloys by EEDM depends on their behavior with respect to the two machining actions. For efficient EEDM, the discharging process dominates the material removal process and is assisted by the dissolution phase of 40 to 60 percent.

The discharges cause thermal depolarization by eroding the oxide layer formed on the anode surface, which, in turn, further enhances the dissolution process. Figure 7.9 shows the relative machinability index that is normalized to 2017 aluminum alloy. This index has been derived on the basis of the VRR at a given machining speed.

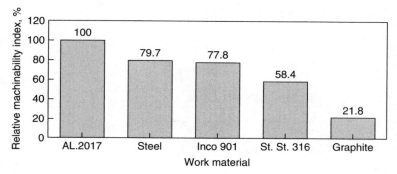

Figure 7.9 Machinability index for some materials normalized to 2017 aluminum (El-Hofy, 1995).

A highly machinable material produces greater removal rates. For a given machining speed, as the removal rate increases, the interelectrode gap becomes wider and machining occurs at a stable mode (Fig. 7.10). Under such circumstances faster rates could be achieved and a higher productivity is also ensured. The low machinability of graphite, limited the machining rates to much lower speeds than those experimented in case of the aluminum alloy (El-Hofy, 1992). Similar to EDM the increase in the removal rate and, hence, machinability is associated with the production of a rough surface. Figure 7.11 shows the shapes of drilled holes by ECM, EDM, and EEDM. Accordingly accurate shapes can be

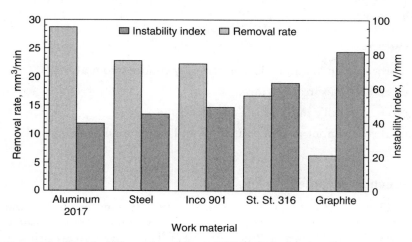

Figure 7.10 Machinability and stability in EEDM (El-Hofy, 1995).

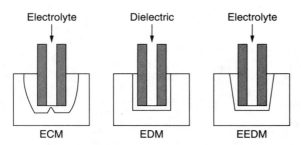
Figure 7.11 Holes drilled by ECM, EDM, and EEDM.

produced by EDM, low-accuracy shapes by ECM, and intermediate-level accuracy shapes by EEDM.

Advantages

- EEDM can produce significantly smoother surfaces due to the presence of high-rate ECD.
- The depth of the heat-affected layer can be significantly reduced or eliminated.
- High machining rates are also possible thereby increasing the productivity and reducing the unit production cost.
- The erosion of tool electrodes is reduced by a factor of 4 to 5 percent compared to that of pure EDM.
- Burrs at the edges are particularly absent due to the existence of the ECD phase.

Disadvantages

- The process can be applied to electrically conductive materials.
- The produced accuracy is low.
- There is difficulty in handling the electrolyte.
- The process cannot produce internal and external sharp edges.

7.3 Electrodischarge Grinding

Electrodischarge grinding (EDG) removes conductive materials by rapid spark discharges between a rotating tool and workpiece that are separated by a flowing dielectric fluid (Fig. 7.12). The spark gap is normally held at 0.013 to 0.075 mm by the servomechanism that controls the motion of the workpiece. The dc power source has capabilities ranging

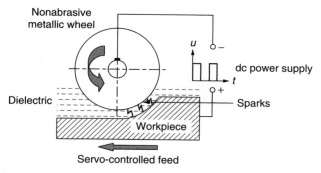

Figure 7.12 EDG schematic.

from 30 to 100 A, 2 to 500 kHz, and 30 to 400 V. The conductive wheel, usually made of graphite, rotates at 30 to 180 m/min in a dielectric bath of filtered hydrocarbon oil. The workpiece is usually connected to the positive terminal of the dc power supply. As can be seen from Fig. 7.12, the workpiece is machined using a stream of electric sparks. Each spark discharge melts or vaporizes a small amount of metal from the workpiece surface. Higher machining currents produce faster rates of machining, rougher finishes, and a deeper heat-affected zone (HAZ) in the workpiece. Less current is used for the production of smoother and less damaged surfaces. Additionally, higher pulse frequencies makes smoother surfaces.

Wheel wear ranges from 100:1 to 0.1:1 with an average of 3:1 depending upon the current density, workpiece material, wheel material, dielectric, and sharpness of corner details. Material removal rates range from 0.16 to 2.54 cm^3/min. Surface finishes in the range of 1.6 to 3.2 μm R_a are possible. Figure 7.13 shows the relationship between removal rate and

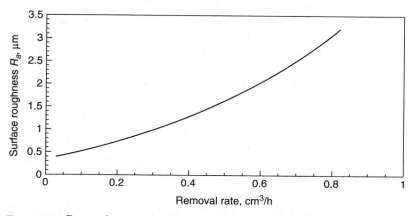

Figure 7.13 Removal rate and surface roughness in EDG (El-Hofy, 1995).

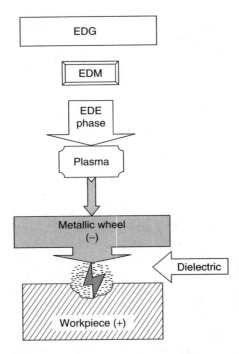

Figure 7.14 Elements of EDG.

surface roughness in EDG. The corner radius depends on the overcut and ranges from 0.013 to 0.130 mm. Greater voltages permit larger gaps, which makes the process suitable for plunge grinding where ease of dielectric flushing is ensured. Tolerances of ±0.005 mm are normal with ±0.001 mm possible. The surface finish improves with an increase in pulse frequency and is typically 0.4 to 0.8 μm R_a. These layers must be removed or modified in case of highly stressed applications. Figure 7.14 shows the main elements of EDG.

Abrasive electrodischarge grinding (AEDG) employs the interactive effect of EDE and MA in order to enhance the machining productivity. In the AEDG process the metallic or graphite electrode used in electrodischarge grinding is replaced by a metallic bond grinding wheel. Therefore, electroerosion in addition to the MA action occurs as shown in Figs. 7.15 and 7.16. An increase in performance measures of the machining process becomes evident when machining superhard materials (plates with synthetic polycrystalline diamond), engineering ceramics, sintered carbides, and metallic composites.

Apart from the above-mentioned effects, the electric discharge causes a considerable decrease in grinding forces, lowers the grinding wheel wear, and provides an effective method for dressing the grinding

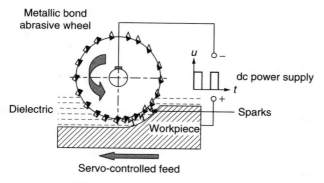

Figure 7.15 AEDG machining system components.

wheel during the machining process. The relative material removal rate for the EDG and AEDG processes was compared to the material removal rate of the electrodischarge process (under the same conditions). Accordingly, the increase in productivity of the EDG process is attributed to improvements in hydrodynamic conditions of dielectric flow. This improvement results from the rotation of the wheel electrode and, partly, by its radial oscillations. Further improvement in EDG performance characteristics can be achieved with the use of vibrating electrodes.

Introducing mechanical effects into the AEDG process (Fig. 7.17) leads to a further increase in the metal removal rate by about 5 times that of the EDM process and about twice that of the EDG process. As the number of wheel revolutions increases, the effect of abrasive action

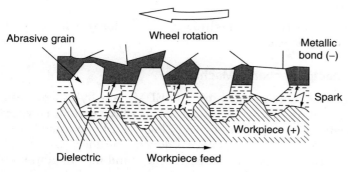

Figure 7.16 Material removal mechanisms in AEDG.

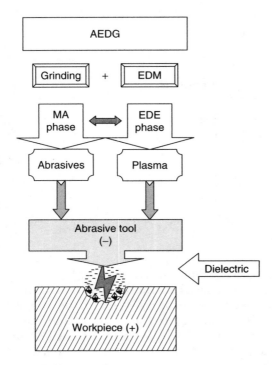

Figure 7.17 Elements of AEDG.

is also increased. This may be evidence of better utilization of electrical discharge energy.

Applications. EDG and AEDG can be used on

1. Steel and carbide at the same time without wheel loading
2. Thin sections on which abrasive wheel pressures might cause distortion
3. Brittle materials or fragile parts on which abrasive materials might cause fracturing
4. Form tools and tungsten carbide throw away bits for which diamond wheel costs would be excessive

7.4 Abrasive Electrodischarge Machining

Abrasive electrical discharge machining (AEDM) is a hybrid process based on EDM, where free abrasive grains, such as silicon carbide powder, are added to the dielectric liquid as shown in Fig. 7.17. In addition to the major EDE thermal phase, MA assistance is added. The energy is carried by the plasma channel in case of the EDE phase and also by abrasive grains for MA action (*www.Jobshop.com/*). The machining media consists of dielectric and solid particles, as indicated in Fig. 7.18. Mixing silicon

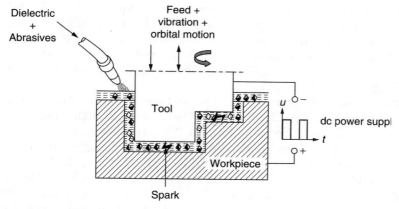

Figure 7.18 AEDM system.

powder into the dielectric reduces the electrical capacitance across the discharge gap by increasing the gap size. As a result, better dispersion of sparks and improvement in the discharge characteristic, especially in the machining of a large workpiece, are ensured by Mohri et al. (1993) and Masui et al. (1995). Figure 7.19 shows the main components of AEDM.

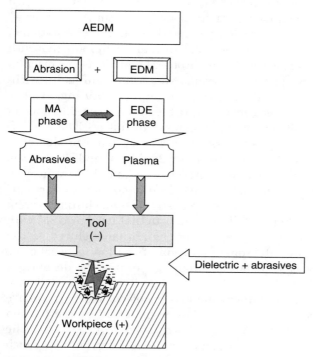

Figure 7.19 AEDM components.

The introduction of mixed powders as working media allows AEDM to produce complex shapes having a mirror finish and an affected layer, which is uniform and free from cracks. AEDM is widely used to produce dies without the need to remove the heat-affected layer using mechanical polishing. The process is, therefore, widely used for producing plastic molding dies.

AEDM is also adopted to modify the surface characteristics of tool steels. The fine tungsten carbide powder, added to the dielectric, adds alloying elements and hence enhances the hardness and corrosion resistance of tools and dies. It also improves their tribological features and changes the surface roughness (Masui et al., 1995). Mohri and coworkers (1993) deposited aluminum on steel and titanium carbide on aluminum using, respectively, aluminum and titanium-based alloy electrodes. Powders such as silicon and tungsten have been used to harden and increase the corrosion resistance of steel workpieces.

7.5 EDM with Ultrasonic Assistance

The need for machining hard and brittle material is steadily increasing in many applications. Traditional machining as well as some of the nonconventional machining processes present several problems related to tool life and produced surface cracks. EDM can be applied to conductive hard and brittle material of electrical resistivity below 100 $\Omega \cdot$cm. A major drawback of the process is the low efficiency of sparking and the formation of open circuit, arc, and short circuit pulses, which are nonproductive with respect to the material removal rate. On the other hand ultrasonic machining can tackle hard and brittle materials, which have difficulty withstanding the cutting forces during conventional machining. At low machining rates, parts having a good surface finish can be produced using USM. The interaction between machining mechanisms of EDM and US vibration in one machining process (EDMUS) has a greater productivity than the sum of the productivities of the individual EDM and USM at the same machining conditions. The new process is adapted for rapid production of graphite electrodes for EDM (*www.Jobshop.com/*) where cutting, drilling, and engraving can be done easily. The machining system for EDMUS is shown in Fig. 7.20. It is similar to that used in USM with the dielectric (deionized water) replacing the abrasive slurry as the machining medium. The ultrasonic vibration of the tool or workpiece together with the low-cost dc power supply generates the discharges across the machining gap.

The material removal mechanism employs the electrodischarge phase for melting and evaporation of the workpiece material, by the plasma channel, in a similar way to normal EDM as shown in Fig. 7.21. The intensity of the discharging action depends on the working conditions such

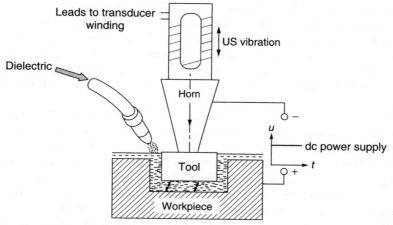

Figure 7.20 EDMUS schematic.

as gap voltage, amplitude, and frequency of ultrasonic vibration as well as the gap flushing conditions. Because of the ultrasonic vibrations of the tool, its front surface moves down toward the workpiece surface and the electric field intensity increases. The electric field, however, causes

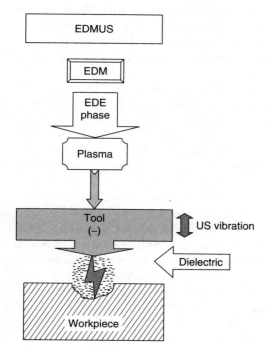

Figure 7.21 EDMUS components.

the ultrafine particles (metal debris ejected from the formed crates) in the machining medium to be suspended and form a bridge across the gap. At a very small gap size, deionization or breakdown of the working fluid occurs and the plasma channel is formed. When the tool front moves up, the gap voltage rises, current drops, and the plasma channel collapses, at large gap sizes. The process begins again when the tool moves down.

Zhixin et al. (1995) used ultrasonic vibrations as a gap flushing method in mechanical pulse electrodischarge machining (MPEDM). Accordingly, cavitation bubbles and the ultrasonic field force prevent the sedimentation of the debris particles in the working fluid. The high-frequency pumping action improves the working fluid circulation by pushing the debris away and sucking fresh fluid into the machining gap. These stoutly increase the discharge efficiency and give higher erosion rates. The flushing action has been reported to increase with vibration amplitudes. Murti and Philip (1987) added that with the application of ultrasonic vibration the machining rate and surface finish improved significantly, and although the tool wear rate increased, the wear ratio was not influenced significantly. Hewedy (1995) reported that the increase of tool vibration up to 100 μm enhanced the erosion rate due to the improved flushing action and hence the sparking efficiency. The presence of too much debris in the electrode gap causes continued arcing and short circuits which makes the process unstable (Masuzawa, et al., 1992). The dielectric must, therefore, contain an appropriate amount of debris since a medium that is too clean is another reason for gap short circuiting and the consequent machining troubles.

The breakdown distance of the electric field is greatly influenced by the contamination level of the machining medium. The gap distance, therefore, changes from one place to another according to the local contamination level. Such a change causes dimensional errors in the copying process. If the concentration is made uniform anywhere in the interelectrode gap, the machining gap becomes uniform and an accurate copy is ensured. Ideal flushing must, therefore, maintain a constant and uniform debris concentration. In this regard, jet flushing with nozzles, sweeping along the interelectrode gap, has been introduced by Masuzawa et al. (1992) and proved to be effective in precision EDM. Schumacher (1990) concluded that the gap contamination through eroded debris influences ignition delay as well as the discharge location in the gap. Erden (1979) investigated the role of dielectric fluid velocity in EDM. Wells and Willey (1975) recommended a dielectric flow velocity of 6 m/s, which reduced the electrode wear by a factor of 2 and the surface roughness by up to 50 percent, increased the machining rate, and reduced the thermal effects in the eroded workpiece surface.

TABLE 7.1 Comparisons between EDM, USM, and EDMUS

Workpiece material	Material removal rate, mm³/min		Surface roughness R_a, μm	
	SG-4 ceramic	SiC	SG-4 ceramic	SiC
USM	1.81	1.95	0.93	0.95
EDM	2.68	2.50	7.90	6.8
EDMUS	5.15	4.95	2.34	2.36

SOURCE: Zhixin et al. (1995).

Table 7.1 shows a comparison between EDM, USM, and the hybrid EDMUS. It can be concluded that the EDMUS material removal rate is about 3 times greater than that of USM and 2 times greater than that of conventional EDM. Moreover the surface roughness of the hybrid process is greatly reduced to one-third of normal EDM by the introduction of US vibration to the tool or workpiece. Surface roughness produced by normal USM is 40 percent of those machined by EDMUS. Zhixin et al. (1995) reported a significant increase in removal rate with applied voltage and little effect to the vibration amplitude. Additionally, the work of Zhang et al. (1997) and Zhang and coworkers (2004) showed also that using EDMUS, the material removal rate and the surface roughness both increase with an increase in voltage, vibration amplitude, and discharge current.

A second beneficial effect of EDMUS concerns the structure modification. The alternate motion of the vibrating electrode, at the ultrasonic frequency, creates more turbulence and cavitations that lead to better ejection of the molten metal from the craters. This in turn enhances the removal rate and reduces the recast material in the machined surface. Consequently, the structure modifications are minimized; less microcracks are observed, which increases the fatigue life of the machined parts if compared to normal EDM.

7.6 Electrochemical Discharge Grinding

Electrochemical discharge grinding (ECDG) combines the electrodischarge erosion (EDE), ECD, and the MA of the grinding process. In the schematic diagram of ECDG, shown in Fig. 7.22, the grinding wheel is connected to the negative terminal while the workpiece is connected to the positive polarity of a pulsed power supply. The electrolyte flows into the interelectrode gap. The rotating wheel is set at a depth of cut, while the workpiece is fed at a constant rate. Surplus material is

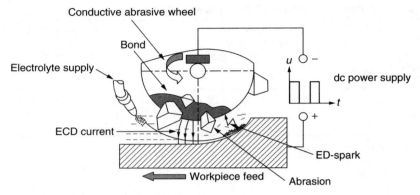

Figure 7.22 ECDG schematic diagram.

removed from the workpiece surface by the anodic dissolution ECD phase, the MA action of abrasives or diamond grains, and the erosion phase due to the spark discharges. Figure 7.23 shows the main material removal mechanisms involved in the ECDG process.

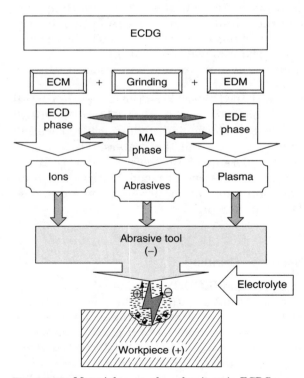

Figure 7.23 Material removal mechanisms in ECDG.

The process occurs due to the application of pulsed voltage in electrolytic solutions. The most common electrolytes are solutions of $NaNO_3$, $NaNO_2$, $NaPO_4$, and KNO_3. It is customary to use ECDG electrolytes with surfactants, corrosion inhibitors, and complex forming agents.

The surfactants enhance the effect of abrasive or diamond grains. Chromates and bichromates are common corrosion inhibitors. The complex forming agents (potassium or sodium salts of tartaric or citric acid) stabilize anodic dissolution because they convert the reaction products into soluble components (Rumyantsev and Davydov, 1984). The gas evolution occurring during the dissolution phase enhances the discharging action and hence the material removal rate. The increase of feed rate and wheel penetration depth enhances the discharge phase contribution in the overall material removal, which, in turn, reduces the overall energy consumption. This is related to the fact that the specific energy consumption is lower in EDM than ECM.

During machining, every discharge causes evaporation of a small metal portion into the electrolyte solution where the solidified metal is flushed away by the electrolyte flow. The higher the pulse energy, the deeper the crater formed, the larger the machining productivity, and the rougher the machined surface. Discharges occur frequently at random locations in the interelectrode gap. Moreover the ECD phase dissolves the rest of the surface electrochemically. Under such conditions the craters formed by the discharge action are smoothed and, consequently, the surface roughness is reduced.

The erosion phase plays a considerable role in the material removal process. Spark discharges cause depassivation of the oxide layer formed on the workpiece surface during ECM, which enhances the dissolution phase. The discharges occurring destroy the glazed layer on the wheel surface; hence, new grains appear which further activates the MA action and the depassivation process.

The machinability of materials by ECDG depends on the properties of the workpiece material under mechanical, electrochemical, and thermal effects. The contribution of each phase is difficult to determine due to the nonuniformity of the wheel electric properties and the variation of the local gaps between the wheel and workpiece. The debris, metal deposits, and gases that exist in the interelectrode gap cause random variations in the gap resistance and hence the current density. The contribution of each machining phase, therefore, depends on the process variables and the instantaneous gap phenomenon under these conditions. Additionally the intensity, location, and duration of each machining phase vary in a random manner that makes the assessment of the material removal rate an even more complicated task. For efficient machining, however, the electrolyte supply should be maintained under pressure. The use of diamond wheels increases the specific removal rate by 20 to 25 percent of that of the metal bonded abrasive wheel during ECDG of 0.14% C–17% Cr–2% Ni steel using 10% $NaNO_3$ solution. However the reduced removal rate and the high wear of the abrasive wheel are more than offset by its low cost. For high machining productivity the electromotive force arising between the two electrodes is used to actuate automatic redressing of the wheel by reversing the polarity (*www.Jobshop.com/* and Rumyantsev and Davydov, 1984).

ECDG produces a high surface finish free from microsize cracks and burrs. The applied voltage affects the height of micro-irregularities, longitudinal feed rate, and grinding depth. The increase in feed rate reduces the height of surface irregularities and speeds up tool wear. In contrast, the increase in the gap voltage results in higher micro-irregularities and tool wear.

7.7 Brush Erosion-Dissolution Mechanical Machining

The brush erosion-dissolution mechanical machining (BEDMM) process is based on machining by ECD, EDE, and friction interaction with metal workpieces, in a water-glass solution, as shown in Figs. 7.24 and 7.25. Electrical arcs melt the workpiece from which particles are subsequently removed by friction between its surface and a rotating brush. As shown in Fig. 7.26, the technique is applicable to the finishing components of complex shapes, which are too hard for conventional machining (Nowicki and Spadlo, 1995).

The use of brush electrodes makes the process cost-effective because it reduces the time needed to machine large components. Proper parameter setting, hot electrode material, and type of power supply generator control the intensity of phases occurring in the machining zone. The

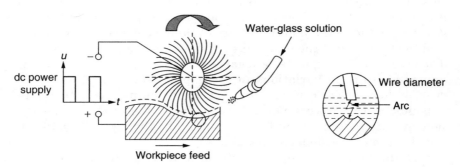

Figure 7.24 Brush mechanical EDM.

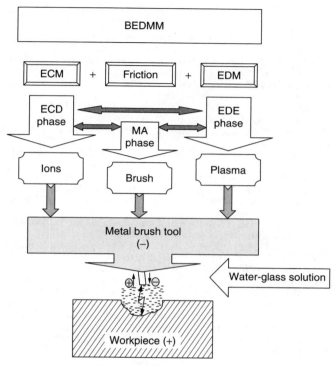

Figure 7.25 Material removal mechanisms in BEDMM.

process allows for both the removal of burrs and flushes and the blunting of sharp edges without damaging or undercutting the machined surface. The selection of electrode material produces a uniform layer, which firmly adheres to the core and has a high content of alloying components (Spadlo, 2002).

Figure 7.26 Parts before and after BEDMM (Nowicki and Spadlo, 1995).

References

Crichton, I., and McGeough, J. A. (1984). "Theoretical, Experimental and Computational Aspects in Electrochemical Arc Machining," *Annals of CIRP*, 33(2):1–4.

El-Hofy, H. (1985). "Fundamental Study of Electrochemical Arc Wire Cutting," Ph.D. Thesis, Aberdeen University, U.K.

El-Hofy, H. (1992). "Electro Erosion Dissolution Machining of Graphite, Inco 901, 2017 Al. and Steels," *5th PEDAC Conf.*, Alexandria, pp. 489–501.

El-Hofy, H. (1995). "Machinability Indices for Some Non Conventional Machining Processes," *Alexandria Engineering Journal*, 34(3): 231–245.

El-Hofy, H. (1996a). "Assessment of Machinability in Electro Erosion Dissolution Machining," *Alexandria Engineering Journal*, 35(2):43–49.

El-Hofy, H. (1996b). "The Role of Electrolyte Flow Velocity in EEDM," *Alexandria Engineering Journal*, 35(2):139–14.

El-Hofy, H., and Khayry, A. (1995). "Comparative Study on Produced Geometry and Surface Integrity by Electrochemical Machining, Electro Discharge Machining, and Electro Erosion Dissolution Machining," *Proc. of IEMS 5*, University of Central Florida, pp. 515–327.

El-Hofy, H., and McGeough, J. A. (1988). "Evaluation of an Apparatus for ECAM Wire Cutting," ASME, *Journal of Engineering for Industry*, 110:119–123.

El-Hofy, H., and McGeough, J. A. (1989). "Effect of Pulse Duty Cycle on Stock Removal Rate and Accuracy in ECDM," *PEDAC-4 Conf.*, Alexandria, pp. 355–362.

El-Hofy, H., El-Makky, M., and El-Wardany, T. (1990). "Surface Generation in Electro Erosion Dissolution Machining," *Alexandria Engineering Journal*, 29(4):301–305.

Erden, A. (1979). "Investigation on the Role of Dielectric Fluid Velocity in EDM," *2nd PEDD Conf.*, Ain Shams University, Cairo, pp. 769–776.

Hewedy, M. (1995). "The Influence of Electrode Vibration in the Performance of EDM Process," *Engineering Research Journal*, Helwan University, Cairo, 2:32–42.

Jain, V. K., et al. (1991). "Experimental Investigations into Traveling Wire Electrochemical Spark Machining (TW-ECSM) of Composites," *Journal of Engineering for Industry*, 113:75–84.

Khayry, A. (1984). "Stochastic and Experimental Studies of Drilling by Electrochemical Arc Machining," Ph.D. Thesis, Aberdeen University, U.K.

Khayry, A., and El-Hofy, H. (1990). "Aspects of Stainless Steel Die Sinking by Electro Erosion Dissolution Machining," *Transactions of SME (NAMRI XVIII)*, pp. 273–243.

Kuneida, M., and Furuoya, S. (1991). "Improvement of EDM Efficiency by Supplying Oxygen Gas into Gap," *Annals of CIRP*, 40(1):215–218.

Levy, M., McGeough, J., and Munro, W. (1983). "Investigation of Surface Finishing of Dies and Moulds in Electrochemical Arc Machining," *ISEM-7*, pp. 265–275.

Machining Data Handbook. (1997). Vol. 2. Cincinnati, OH: Machinability Data Center, Institute of Advanced Manufacturing Sciences.

Masui, K., Demizu, K., Sato, Y., and Sone, T. (1995). "Surface Modification of Tool Steels by Alloying Method Using EDM Processes," *ISEM-11*, pp. 419–426.

Masuzawa, T., et al. (1992). "Improved Jet Flushing for EDM," *Annals of CIRP*, 41(1):239–242.

McGeough, J., Khayry, A., and Munro, W. (1983). "Theoretical and Experimental Investigations of the Relative Effects of Spark Erosion and Electrochemical Dissolution in Electrochemical Arc Machining," *Annals of CIRP*, 32(1):113–118.

Mohri, M., Saito, N., and Tsunekawa, Y. (1993). "Metal Surface Modification by EDM with Composite Electrode," *Annals of CIRP*, 42(1):219–222.

Murti, V. S., and Philip, P. K. (1987). "A Comparative Analysis of Machining Characteristics in Ultrasonic Assisted EDM by Response Surface Methodology," *Int. Journal of Product Research*, 25(2):259–272.

Nowicki, B., and Spadlo, S. (1995). "Brush Electrodischarge Mechanical Machining," *ISEM-11*, pp. 209–216.

Rajurkar, K. P., Wei, B., Kozak, J., and Nooka, S. R. (1995). "Abrasive Electrodischarge Grinding of Advanced Materials," *ISEM-11*, pp. 863–869.

Rumyantsev, E., and Davydov, A. (1984). *Electrochemical Machining of Metals*, Moscow: Mir Publishers.

Saushkin, B., et al. (1982). "Special Features of Combined Electrochemical and Electro Erosion Machining of Elongate Machine Parts," *Electrochemistry in Industrial Processing and Biology* 105, part 3, 8–14.

Schumacher, B. (1990). "About the Role of Debris in the Gap During EDM," *Annals of CIRP*, 39(1):197–199.

Spadlo, S. (2002). "Computer Studies of Brush Electrodischarge Machining with Electrode of Alloy Steel and Tungsten," *16th Int. Conf. on Advanced Manufacturing Systems and Technology Proceedings* (AMST03), Udine, Italy, pp. 515–524.

Thermal Machining Processes. (1979). Dearborn, MI: SME.

Wells, P., and Willey, P. (1975). "The Effects of Variation in Dielectric Flow Rate in the Gap on Wear Ratio and Surface Finish During EDM," *IEE Conf. on Electrical Methods of Machining, Forming and Cutting*, pp. 110–117.

www.Jobshop.com/techinfo/electdischagrggrindingdef.shtml.

www.unl.edu/nmrc/hybrid/hybrid_machining.htm.

Zhang, J. A., Lee, T. C., Lau, W. S., and Ai, X. (1997). "Spark Erosion with Ultrasonic Frequency," *Journal of Material Processing Technology*, 68:83–88.

Zhang, Q. H., Zhang. J. H., Ren, S. F., Deng, J. X., and Ai, X. (2004). "Study on Technology of Ultrasonic Vibrations Aided Electrical Discharge Machining in Gas," *ISEM XIV On-site Conf. Proc.*, Edinburgh, U.K. (*Journal of Materials Processing Technology*).

Zhixin, J., et al. (1995). "Study on Mechanical Pulse Electrodischarge Machining," *Precision Engineering*, 17(2):89–93.

Chapter 8

Material Addition Processes

8.1 Introduction

Rapid prototyping (RP) is a group of modern manufacturing technologies that are used to produce three-dimensional prototypes from CAD representations. These methods are generally similar to each other in that they add and bond materials in a layered fashion to form objects. These techniques are also referred to as solid free-form fabrication (SFF). RP applies similar physical and chemical phenomena to progressively add materials as those used for selective material removal during the so-called nontraditional machining processes, dealt with in the previous chapters and summarized in Table 8.1. Most of the RP processes are still in the precommercial development stage. While object sizes range from microscopic to entire buildings, materials range from paper and plastic to metals and ceramics. Applications include toys, aerospace, and advanced medicines. One of the success factors of RP is related to its ability to produce complex net and near net shaped parts in materials that are hard to machine conventionally, like difficult-to-machine metals, ceramics, and composites.

Advantages of rapid prototyping

- Builds complex three-dimensional geometrical shapes
- Is automated based on CAD models
- Uses a generic fabrication machine that does not require part specific tooling
- Requires a minimum of or no human intervention
- Produces accurate prototypes in a short time at a minimum cost

TABLE 8.1 Physical and Chemical Phenomena on Material Removal and Material Addition Processes

Phenomena	Material removal	Material addition
Chemical processes	CHM ECM	Stereolithography (SLA) Photopolymerization Laser-induced carbon vapor deposition (CVD)
Thermal processes	LBM PBM EBM EDM	Selective laser sintering (SLS) Plasma spraying* Electron beam (EB) sintering* Electrodischarge (ED) deposition*
Liquid jet	WJM	3-D ink-jet printing
Solid jet	AJM	Powder jet laser cladding
Ultrasonic processes	USM	N/A

*Not applied commercially.
SOURCE: Kruth et al. (1998).

Classification of RP techniques. Kruth (1991) classified the RP techniques according to the raw material used into

1. Liquid-based techniques
2. Powder-based techniques
3. Solid-based processes

Some of these methods depend on solidifying a liquid polymer by the impact of light or laser or by the curing of powders to create the prototype. Other processes are based on melting, deposition, and resolidification of plastic or resin materials.

8.2 Liquid-Based Techniques

8.2.1 Stereolithography

Stereolithography (SLA) is the most common RP technique that employs layer-by-layer manufacturing based on photopolymerization. Solidification happens by photopolymerization that results from the impact of a laser light on the upper surface of the liquid. Light polymerization of the liquid is limited to a layer thickness of a few tenths of a millimeter below the surface. Illumination and hence polymerization of the surface liquid is restricted to a pattern that corresponds to the part's cross section. Once the layer is completely traced, it is lowered by a small distance into the liquid and a second layer is traced right on top of the first. The self-adhesive property of the material causes the layers to bond to one another and eventually form a complete, three-dimensional object after many such layers are formed. Some objects have overhangs or undercuts, which must be supported during the fabrication process by support structures. These are either manually or automatically designed or fabricated right along with the object. Upon completion of the fabrication process, the object is elevated from the liquid and the supports are cut off (*home.att.net/~castleisland/sla*).

The following is the sequence of the SLA process:

1. Slicing software reads the CAD model and cuts it into thin slices that depend on the resolution of the mechanical movements of the RP system as well as on the part geometry.
2. The computer receives the output from the slicing software that controls the corresponding transitions and rotations of the mechanical components of the RP system.
3. The process chamber filled with the photosensitive liquid polymer has an elevator-like platform that can be raised vertically or lowered.
4. The laser beam scans the selected surface area of the photopolymer. It cures the scanned patterns and produces a solid section.
5. The platform is then lowered so that another shallow layer of the liquid polymer covers the previously cured layer.
6. The sequence is repeated to produce the complete part geometry.

Controlling the laser beam and the platform movements through the servo-controlled system (Fig. 8.1) can therefore produce a variety of parts.

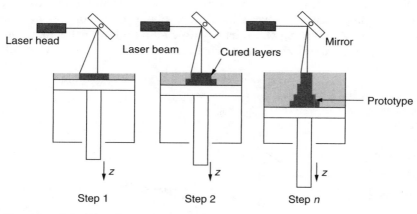

Figure 8.1 Principles of stereolithography.

Advantages of SLA

- It has a high accuracy of ±0.1 mm and produces a good surface finish.
- It produces semitransparent materials that can be processed for optical clarity.
- The process is fully automated.

Disadvantages of SLA

- There is a limited range of materials that can be used.
- It has relatively high cost compared to other RP methods.
- Working with liquids can be messy.
- Parts often require postcuring operation.

8.2.2 Holographic interference solidification

Holographic interference solidification (HIS) is based on photopolymerization. The idea is that projecting a holographic image of a model in a vat of liquid photosensitive monomer could solidify a whole three-dimensional surface at once, rather than to build it up point by point. This process saves the prototyping time. Reported applications include the production of lost wax models for copper EDM electrodes.

8.2.3 Beam interference solidification

Beam interference solidification (BIS) is the most versatile polymer-based RP process. In beam interference solidification (Fig. 8.2) part creation occurs by point-by-point solidification of the liquid at the intersection of two laser beams having different frequencies. All the liquid hit by laser beam 1 is excited so that it polymerizes upon the impact of laser beam 2. Forming the part in the z axis no longer needs to be done in layers. Initializing the laser beam movements in any set of

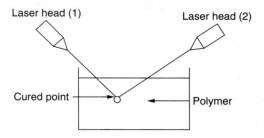

Figure 8.2 Beam interference solidification.

Material Addition Processes 233

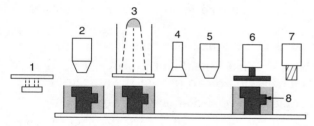

1. Mask preparation
2. Photopolymer resin application
3. UV light exposure
4. Aerodynamic wiper
5. Wax application
6. Wax cooling
7. Milling
8. Prototype

Figure 8.3 Solid ground curing steps.

x, y, z coordinates makes it possible to trace the part in three dimensions rather than in two in a small prototyping time (*mtiac.alionscience.com*). The process has a limited industrial application because of the following problems:

1. Light absorption (drop of intensity with depth)
2. Shade effects of parts already solidified
3. Beam intersection problems due to the light diffraction caused by local temperature variation or solidification

8.2.4 Solid ground curing

A schematic diagram for solid ground curing (SGC) is shown in Fig. 8.3. Accordingly, the process has two main cycles, the photomask generation cycle and a layer fabrication cycle. It takes about 2 min to complete all operations to make a single layer. The following steps are followed during SGC (*claymore.engineer.gvsu.edu*):

1. CAD files are sliced.
2. A layer of photosensitive polymer is first sprayed onto the platform.
3. A photomask is prepared (for each layer) using electrophotography (xerography).
4. The mask plate is then placed on top of the platform and object under construction.
5. Ultraviolet light is exposed onto the layer, which solidifies completely.

6. The residual polymer is vacuumed for recycling and replaced by wax.
7. The wax is hardened using the wax cooling plate.
8. The milling head is used to machine the photopolymer to an accurate and uniform thickness in preparation for the next layer.
9. The mask plate is erased and reused after every layer.
10. The process is repeated for the next slice until the object is completely finished.

Secondary operations are required to remove the wax, which can either be melted away or dissolved before the object is finished. The wax matrix makes it unnecessary to generate extra support structures for overhangs or undercuts (*home.att.net/~castleisland/sgc*).

Advantages of SGS

- Complicated and large components are produced.
- There is no need for additional supports.
- No postcuring is needed, which reduces internal stresses and warping.
- Multiple parts can be run at the same time due to the large workspace.
- Accuracy is good.
- Part complexity does not affect speed; however, the volume does.
- Weights may be inserted at any time to alter the center of gravity.
- Layers can be milled off if they are found to be in error.

Disadvantages of SGS

- The cost of equipment is high.
- The machine size is large.
- The process is rather complicated and requires high maintenance and skilled supervision.
- The range of materials is limited.
- The wax must be removed after production in a large oven.
- The process is noisy.
- Overexposure of the polymer may increase the viscosity and make it unusable, thus greatly increasing the volume of expensive polymers used.
- The resins require that light sealed chambers and toxic material handling procedures be used.

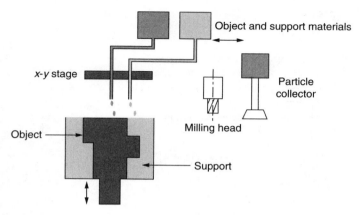

Figure 8.4 Liquid thermal polymerization.

8.2.5 Liquid thermal polymerization

Liquid thermal polymerization (LTP) uses thermoset instead of photopolymer, which is normally used in stereolithography. Solidification occurs by heat dissipation rather than by the impact of laser light. Heat dissipation might make it difficult to control the accuracy, thermal shrinkage, and part distortion. The system employs two jets for the plastic object and the waxlike support materials. The liquids are fed to the individual jetting heads, which squirt tiny droplets of the materials as they are moved in x-y coordinates to form a layer of the object as shown in Fig. 8.4. The materials harden by a rapid drop in temperature as they are deposited.

After an entire layer of the object is formed by jetting, a milling head is passed over the layer to make it a uniform thickness. As the milling head cuts, particles are vacuumed and collected. The process is repeated to form the entire object. The wax support material is either melted or dissolved away. The most outstanding characteristic of LTP is the ability to produce extremely fine resolution and surface finishes that are equivalent to CNC machining. However, the technique is very slow for large objects. While the size of the machine and materials used are office-friendly, the use of a milling head creates noise which may be objectionable in an office environment. Materials selection also is very limited (*home.att.net/~castleisland/ink*).

8.2.6 Fused deposition modeling

Fused deposition modeling (FDM) is the second most widely used RP technology, after stereolithography. The process is unlike all the other RP processes because a laser is not used to produce the cross-sectional layer.

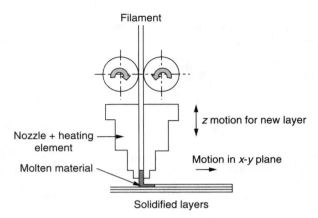

Figure 8.5 FDM principles.

Instead, the model material is applied directly in a semiliquid form from a CNC controlled extruder head. FDM is the favored type of RP process when tough plastic components are needed. The process depends on building up the model layer-by-layer using a thermoplastic 0.003-mm-diameter filament. The filament is unwound from a coil and supplies material to an extrusion nozzle as shown in Fig. 8.5. The filament melts as it flows through the heated extrusion head and emerges as a thin ribbon through an exit nozzle. The FDM system operating software imparts a CAD file. The software then orients the part for optimum building, slices the model, and creates the path generated by the operating software, laying down a very thin layer of molten material. One layer upon another, the model is built from the bottom up. The required object is produced as a three-dimensional solid part in a single stage without need for tooling. Each formed surface layer is machined with a milling cutter prior to the application of the next layer in order to control the flow of polymer, and an evenly distributed layer is obtained. Recent FDM systems have two extrusion heads: one for modeling material and the other for modeling support structure material (*www.quickmodels.com* and *www.caip.rutgers.edu*).

In the apparatus shown in Fig. 8.6 the nozzle is mounted to a mechanical stage, which can be guided around by CNC code and the viscous ribbon of polymer is gradually built up from a fixtureless base plate. As the nozzle is moved over the table in the required x, y, z geometry, it deposits a thin bead of extruded plastic to form each layer. The plastic hardens immediately after being squirted from the nozzle and bonds to the layer below. The entire system is contained within a chamber, which is held at a temperature just below the melting point of the plastic material.

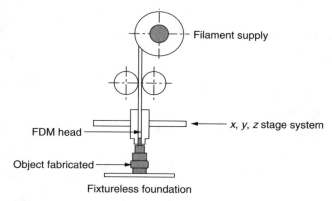

Figure 8.6 Fused deposition modeling schematic.

A second nozzle may extrude a second material if required. When the first layer is complete, the nozzle moves upward by one layer thickness and the process begins again. The part is easily removed from the platform, supports are dissolved, and the part becomes ready. Many parameters are involved in building a part such as the basic part design: the material selection, tool path, layer thickness, and part orientation. The time to build the part, cost, strength, and accuracy of the part depend on these parameters.

Several materials are available for the process including acrylonitrile butadiene styrene (ABS), medical grade ABS, E20 (a thermoplastic polyester-based elastomer), and investment casting wax. ABS offers good strength, and more recently polycarbonate and poly(phenyl) sulfone materials have been introduced, which extend the capabilities of the method further in terms of strength and temperature range. For simple parts there is no need for fixturing and material can be built up layer by layer. For complex parts with inner cavities, unusual sculptured surfaces, and overhanging features, it is sometimes necessary to create support structures for any overhanging sections of the model. Support structures are later removed by breaking them away from the object. A water-soluble support material that can simply be washed away is also available (*students.bath.ac.uk*).

Advantages of FDM

- Quiet, nontoxic, and office-friendly process
- Fast for small, thin shapes
- Produces strong parts
- Cost-effective

238 Chapter Eight

- No need for part cleanup
- Able to produce multicolored parts using colored ABS

Disadvantages of FDM

- Slow when making large cross-sectional areas
- Has to use additional support material
- Relatively low accuracy and poor surface finish
- Unable to make complicated parts
- Limited number of materials available

8.2.7 Multijet modeling

In multijet modeling (MJM) a printing head comprising 96 jets oriented in a linear array builds models in successive layers. Each jet applies a specially developed thermopolymer material only where necessary. As shown in Fig. 8.7, the MJM head shuttles back and forth like a line printer (x axis), building a single layer of what will soon be a three-dimensional model. If the part is wider than the MJM head, the platform repositions (y axis) to continue building the next layer. This process is continued until the model is complete. The large number of jets allows fast and continuous material deposition for maximum efficiency using inexpensive thermopolymer (*www.foundryonline.com*).

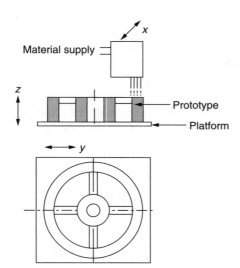

Figure 8.7 Multijet modeling principles.

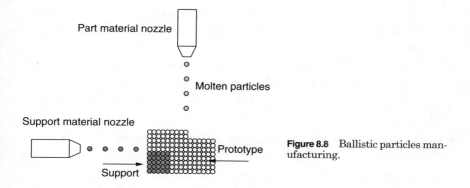

Figure 8.8 Ballistic particles manufacturing.

8.2.8 Ballistic particles manufacturing

In ballistic particles manufacturing (BPM), a stream of molten material is ejected from a nozzle. The material separates into droplets, which hit the substrate and immediately cold weld to form a part (Fig. 8.8). The material stream may be a drop-on-demand system or continuous jet using a piezoelectric transducer that excites the nozzle at 60 Hz. The nozzle excitation forces ensure the production of a stream of small, regular droplets with uniform spacing and distance (*ime.ucla.edu*).

PBM parameters include temperature and velocity of the droplets and the electronic charge they require when the stream is ejected, which is used to guide the droplets to the surface. The temperature controls the speed at which the molten metal droplets solidify. Furthermore, the deformation and placement accuracy of the droplet depends on its velocity.

This process uses CAD-generated three-dimensional solid model data to direct the stream of material (waxes, plastics, photocurable polymers, ceramics, or metals) at a target, building three-dimensional objects in much the same manner that an ink-jet printer produces two-dimensional images. The object is built using a three-axis robotic system controlling a piezoelectric ink-jet mechanism for shooting particles of the material onto a target and producing multiple cross sections. A second material supports any overhanging structures on the part. When the product is complete, it is placed in a bath that dissolves the support material. A limitation here is that the support material cannot be completely enclosed by the construction wax (Hauber, 1988).

Advantages of BPM

- No postproduction finishing is required.
- Tolerances of ±0.01 mm can be achieved.

- The ability to perform in microgravity and vacuum environments provides the production of parts in space and for advanced materials research.
- The method can however be extended to other materials that easily melt and solidify such as thermoplastics and metals.
- Wax models for investment casting can be produced without the need for dies.
- It allows the application of different materials or colors within a single part.
- Decreased turnaround time, increased productivity, and improved material properties make the BPM technology economically attractive.
- The method produces rapidly solidified microstructures in situ.
- Near full density and net shape parts in a unit operation are possible.

8.2.9 Shape deposition manufacturing

Shape deposition manufacturing (SDM) is a new fabrication methodology for automatically building up material layers to form three-dimensional, complex-shaped, multimaterial structures. The process combines the advantages of solid free-form fabrication (i.e., easy to plan and does not require special fixturing, arbitrarily complex shapes, and heterogeneous structures) with the advantages of conventional machining (i.e., high accuracy, good surface finish, and wide-scale availability of existing three- and five-axis CNC milling machines). In addition to the rapid prototyping of complex shapes, selective additive material processing enables the fabrication of multimaterial structures and permits prefabricated components to be embedded within the growing shapes. The goal of SDM is, therefore, to automatically manufacture parts that are strong, geometrically accurate, and have a good surface finish (*www-2.cs.cmu.edu*). Figure 8.9 shows the manufacturing steps, which are as follows:

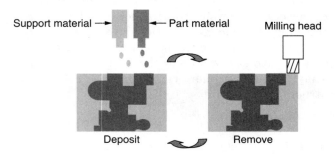

Figure 8.9 Shape deposition manufacturing.

1. A CAD model is first decomposed into thin layer representations.
2. A layered segment is deposited as near-net shapes. The thickness of that layer depends on the local part geometry and the deposition process constraints.
3. A layer of sacrificial support structure is simultaneously built up to support and create a fixture for the growing shape.
4. The layers are machined to a net shape before additional material is deposited.
5. The sequence for depositing and machining the part and support materials is repeated to form the entire shape.
6. After the entire part is built up, the sacrificial support material is removed to reveal the final part.

8.3 Powder-Based Processes

8.3.1 Selective laser sintering

Selective laser sintering (SLS) is a free-form fabrication method based on sintering of the metallic or nonmetallic powders selectively into an individual object. As shown in Fig. 8.10, a thin layer of fusible powder is laid down and heated to just below its melting point by infrared heating panels at the side of the chamber. Then a laser beam guided on the basis of a three-dimensional CAD system of the part to be produced, sinters and fuses the desired pattern of the first layer of the object in the powder.

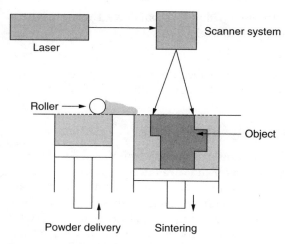

Figure 8.10 Selective laser sintering.

Next this first fused layer descends, the roller spreads out another layer of powder, and the process repeats. SLS relies on the supporting strength of the powder around the partially fused object. Therefore, support columns for any overhanging parts of the component are not needed. A variety of materials can be used including polymers, epoxy, wax, metals, and ceramic with the appropriate binder. After the object is fully formed, the piston is raised to elevate it. Excess powder is simply brushed away and a final manual finishing may be carried out. Although surface finishes and accuracy are not quite as good as with stereolithography, the material properties can be quite close to those of the intrinsic ones. The method has also been extended to provide direct fabrication of metal and ceramic objects and tools (*home.att.net/~castleisland/SLS*).

Advantages of SLS

- It produces the toughest components compared to other RP techniques.
- A wide range of materials can be used, including metal, plastic, ceramics, wax, nylon, elastomers, and polycarbornate.
- A relatively short production time (approximately 1 in/h) is possible.
- No postcuring processes are needed.
- No additional support is required during the building process.
- Leftover powder can be reused.
- Different parts can be built right on top of others.

Disadvantages of SLS

- A rough surface finish is produced due to the stair-step effect.
- The time needed to heat up the powder and cool down the model after it has been completed is approximately 8 to 10 h.
- Toxic gases are produced especially when plastic materials are fused.
- Components are less detailed than those produced by stereolithography due to a high processing temperature and high shrinkage.

8.3.2 Laser engineered net shaping

As shown in Fig. 8.11, a high-power laser is used to melt metal powder supplied coaxially to the focus of the laser beam through a deposition head. The x-y table is moved in raster fashion to fabricate each layer of the object. Typically the head is moved up vertically as each layer is completed. The laser beam is delivered to the work by a simple right-angle mirror. Metal powders are delivered and distributed around the circumference of the head either by gravity or by using an inert, pressurized carrier gas, which

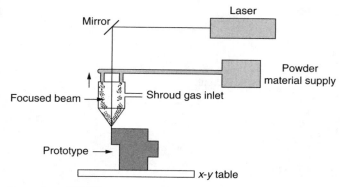

Figure 8.11 LENS process.

can also be used to shield the melt pool from atmospheric oxygen. It also provides better control of properties and promotes layer-to-layer adhesion through better surface wetting (*home.att.net/~castleisland/len*).

A variety of materials can be used such as stainless steel, Inconel, copper, and aluminum. Of particular interest are reactive materials such as titanium. Most systems use powder materials, but there has also been work done with material provided as fine wires. In this case the material is fed off-axis to the laser beam.

The building area is usually contained within a chamber both to isolate the process from the ambient surroundings and to shield the operators from possible exposure to fine powders and the laser beam. The laser power used varies greatly, from a few hundred watts to 20 kW or more, depending on the particular material, feed rate, and other parameters.

Objects fabricated are near net shape, but generally will require finish machining. They are fully dense with good grain structure and have properties similar to, or even better than, the intrinsic materials. Laser engineered net shaping (LENS) has fewer material limitations than SLS, doesn't require secondary firing operations as some of those processes do, and can also be used to repair parts as well as fabricate them. Initial applications are concentrated on the fabrication and repair of injection molding tools and the fabrication of large titanium and other exotic metal parts for aerospace applications.

8.3.3 Three-dimensional printing

Three-dimensional printing combines the features of SLS and BPM, (Fig. 8.12). Powder material is deposited in successive layers, as in SLS. The process starts by depositing a layer of powder object material at the top of a fabrication chamber. To accomplish this, a measured quantity of powder is first dispensed from a similar supply chamber by a piston moving

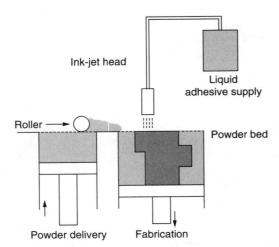

Figure 8.12 Three-dimensional printing.

upward incrementally. The roller then distributes and compresses the powder at the top of the fabrication chamber. The multichannel jetting head subsequently deposits a liquid adhesive in a two-dimensional pattern onto the layer of the powder, which becomes bonded in the areas where the adhesive is deposited to form a layer of the object.

Once a layer is completed, the fabrication piston moves down by the thickness of a layer, and the process is repeated until the entire object is formed within the powder bed. After completion, the object is elevated and the extra powder brushed away leaving the object. No external supports are required during fabrication since the powder bed supports overhangs. After building up the whole part, a heat treatment is applied by curing at 120°C for 2 h. In case of ceramic parts, a final firing at 1000 to 1500°C is required to give the part its full mechanical and refractory strength.

Three-dimensional printing offers the advantages of speedy fabrication and low materials cost. In fact, it's probably the fastest of all RP methods. Recently color output has also become available. However, there are limitations on resolution, surface finish, part fragility, and available materials. The closest competitor to this process is probably fused deposition modeling (FDM).

8.4 Solid-Based Techniques

8.4.1 Solid foil polymerization

This process applies solid-to-solid, rather than liquid-to-solid, polymerization. The part is built up using semipolymerized plastic foils progressively stacked on top of each other. Upon exposure to ultraviolet light, the semipolymerized foil solidifies and bonds to the previous layer. The areas of the foil that do not constitute the eventual part are used to support it during the building process but remain soluble and so are easy to remove once the part is complete.

8.4.2 Laminated object modeling

In laminated object modeling (Fig. 8.13) profiles of object cross sections are cut from paper or other web material using a laser. The paper is unwound from a feed roll onto the stack and first bonded to the previous layer using a heated roller which melts a plastic coating on the bottom side of the paper. The profiles are then traced by an optics system that is mounted to an x-y stage.

After the laser has cut each profile, the roll of paper is advanced, a new layer is glued onto the stack, and the process is repeated. After fabrication, some trimming, hand finishing, and curing are needed. The method is self-supporting for overhangs and undercuts. Areas of cross sections, which are to be removed in the final object are heavily crosshatched with the laser to facilitate removal (*home.att.net/~castleisland/pmkt/lom*).

In LOM the surface finish, accuracy, and stability of paper objects are not as good as for materials used with other RP methods. However, material cost is very low, and objects have the look and feel of wood and can be worked and finished in the same manner to form patterns for sand castings. While there are limitations on materials, work has been done with plastics, composites, ceramics, and metals. Some of these materials are available on a limited commercial basis.

In summary, there are several manufacturing processes available that make important transitions from CAD to a prototype part. Some of theses

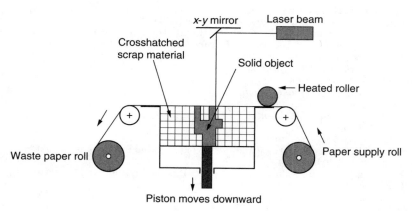

Figure 8.13 Laminated object modeling (LOM).

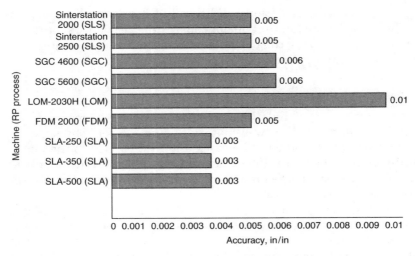

Figure 8.14 Accuracy levels for RP techniques (Wright, 2001).

are in daily commercial use such as SLA, SLS, LOM, and FDM, while more are still in the research and development stage like three-dimensional printing, SGC, and SDM. Currently SLA is the most used RP technique especially for generation of the master patterns for casting and injection molding (McGeough, 2002). After SLA and SLS, FDM, and LOM have the most validity. Figure 8.14 shows the accuracy levels for the different RP techniques.

References

Elsabaa, I., Soliman, E., and Khayry, A. (2001). "A Comparative Study of Rapid Prototyping Techniques," *Alexandria Engineering Journal*, 40(3):335–346.
Hauber, D. (1988). "Automatic Production of PBM Parts Directly from a Computer Aided Design Model," *International Journal of Powder Metallurgy*, 24(4):337–342.
home.att.net/~castleisland/fdm_int.htm
home.att.net/~castleisland/ink_int.htm
home.att.net/~castleisland/len_int.htm
home.att.net/~castleisland/pmkt/lom_int.htm
home.att.net/~castleisland/sgc.htm
home.att.net/~castleisland/sla_int.htm
home.att.net/~castleisland/SLS_int.htm
ime.ucla.edu/~luoh/bpm-web.htm
Kruth, J. P. (1991). "Material Increase Manufacturing by Rapid Prototyping Techniques." *Annals of CIRP*, 40(2):603–614.
Kruth, J. P., Leu, M. C., and Nakagawa, T. (1998). "Progress in Additive Manufacturing and Rapid Prototyping," *Annals of CIRP*, 47(2):525–540.
McGeough, J. A. (2002). *Micromachining of Engineering Materials*. New York: Marcel Dekker, Inc.
mtiac.alionscience.com/pubs/rp/rp42.htm

mtiac.alionscience.com/pubs/rp/rp47.htm
Sachs, E., et al. (1993). "Three-dimensional Printing: the Physics and Implications of Additive Manufacturing," *Annals of CIRP*, 40(1):257–260.
students.bath.ac.uk/en2hlt/FDM2.htm
Wright, P. K., (2001). *21st Century Manufacturing*. Upper Saddle River, NJ: Prentice Hall.
claymore.engineer.gvsu.edu/eod/manufact/manufact-310.html
www.bsu.edu/classes/flowers2/lom1.htm
www.caip.rutgers.edu/~kbhiggin/VDF/FDM.html#intro
www-2.cs.cmu.edu/~sdm/
www.foundry on line.com
www.quickmodels.com/prod03.htm

Index

Abrasive machining:
  conglomerates, 48, 49
  electrodischarge machining, 203, 216
  flow machining, 109
  grain, 6, 15, 21, 22, 24, 25, 39, 49, 50, 184–186, 194, 198, 215, 216
  grit, 29, 183, 189
  jet machining, 34, 49
  slurry, 9, 16, 17, 21, 24–28, 109, 218
  water jet machining, 9, 43, 54, 145
Accuracy:
  of ECM, 90
  levels, 26, 175, 246
  of machined parts, 26, 90
Air plasma, 167, 168, 169, 180
Anodic dissolution, 12, 71, 72, 74, 81, 101, 102, 222, 223
Anodic film, 72
Anodic workpiece, 77, 78, 102, 103, 106, 108, 182, 183, 185
Aspect ratio, 104, 146, 148, 154, 155

Back pressure, 83, 101
Ballistic particles manufacturing, 239, 243
Beam interference solidification, 232, 248
Brass, 15, 23, 54, 73, 84, 121, 123, 144, 153
Brush erosion-dissolution mechanical machining, 224

Capillary drilling, 105, 108, 109
Cast iron, 82, 95
Cathode, 71, 72, 78, 79, 81, 83, 99, 103, 116–118, 152, 160, 166, 172, 173, 182, 196
Cathodic reactions, 77

Cathodic tool, 78, 83, 87, 93, 100, 102, 103, 105, 106, 183, 189, 190, 192, 198
Ceramics, 6, 8, 35, 39, 41–43, 52, 65, 67, 131, 133, 134, 140, 144, 156,177–179, 197, 201, 214, 229, 239, 242, 245
Chemical blanking, 10, 66
Chemical dissolution, 10–12, 55, 62, 181, 182
Chemical etching, 176, 181
Chemical milling, 10, 55, 56, 66
Chemical processes, 12, 55
Chromium, 21, 37, 144, 150, 162, 175
Composites, 8, 43, 133, 149, 150, 178, 188, 201, 204, 214, 226, 229, 245
Computer-aided design, 46, 66
Computer-integrated manufacturing, 83, 130, 138
Computer keyboard, 154, 179
Computer numerical control, 3, 31, 132, 171
Constant current, 90, 91, 96, 97, 112
Constant feed, 80, 85, 86, 96
Constant gap, 85, 96, 119, 120
Continuous wave, 140
Contour machining, 29, 30, 53
Copper, 21, 38–40, 58, 64, 67, 71, 73, 82, 84, 88, 111, 121, 123, 126, 135, 136, 144, 153, 155, 166, 168, 169, 176, 186, 187, 232, 243
Coring, 28
Crater, 27, 118, 125, 126–128, 134, 151, 152, 203, 206, 208–210, 221, 223
Cubic boron nitride, 133
Current:
  density, 71, 72, 73, 82, 84, 88–95, 118, 175, 181, 183, 186, 187, 189, 194–196, 199, 200, 208, 209, 213, 223

249

Index

Current: (*Conti.*)
  efficiency, 82–85, 89, 90, 102, 200
  flow, 96, 97, 112, 118, 196
  lines, 185, 194
Cutting:
  of printed circuit boards, 38
  rate, 34–36, 41, 45, 46, 146, 168, 169
  of rocks, 37
  speed, 43, 121, 143, 149, 153, 166, 171, 172

DC power supply, 72, 81, 183, 193, 195, 198, 205, 213, 215, 217–219, 222, 224
Deburring, 37, 40, 42, 44, 52, 74, 79, 97, 99, 109–112, 172
Dielectric fluid, 118, 121, 123, 124, 212, 220, 226
Dielectric flushing, 124, 131, 203, 214
Dielectric liquid, 103, 203
Dielectric type, 134
Die sinking, 91, 99, 115, 131, 132, 139, 140, 226
Dies and molds, 128, 131, 204
Dissolution phase, 10, 181, 194, 195, 197–199, 203, 210, 223
Dressing of grinding wheels, 152, 179
Dynamic balancing, 155

ECD phase, 87, 184, 189, 191, 193, 200, 203, 212, 222, 223
ECG process, 184–186, 188, 201
ECM:
  accuracy, 92
  equipment, 79
  process, 87, 97
EDE phase, 12, 116, 203, 207, 209, 214, 216, 217, 219, 222, 225
EDM:
  automation, 138
  dielectric, 120, 123
  electrodes, 30, 121, 232
  milling, 131–133
  with ultrasonic assistance, 218
EEDM wire cutting, 205
Electrochemical processes:
  arc machining, 203, 226
  buffing, 12, 182, 196, 201
  deburring, 109, 111
  discharge grinding, 13, 203, 221
  discharge machining, 203
  dissolution, 10, 11, 27, 100, 181, 183, 184, 186, 196, 198, 201, 203, 221, 224, 226
  drilling, 91, 100, 145

  grinding, 11, 181, 182, 184
  honing, 181, 189, 201
  jet drilling, 108
  machining, 10–12, 75, 77, 78, 88, 97, 112, 113, 145,181, 182, 197, 199, 201, 204, 208, 226
  superfinishing, 12, 181, 192
Electrode:
  material, 121–123, 132, 224, 225
  polarity, 121, 125, 134
  wear, 121, 122, 131, 220
Electrodischarge:
  erosion, 12, 115, 203, 206, 221, 224
  grinding, 136, 203, 212, 214, 226
  machining, 9, 10, 15, 130, 145, 177, 203, 204, 216,220, 227
  texturing, 128, 134
Electroerosion dissolution machining, 203, 204
Electrolysis, 77, 79, 101, 102, 110, 206, 209
Electrolyte:
  concentration, 82, 90, 95
  feeding, 79, 83, 96, 101
  flow rate, 90, 92, 95, 101, 196, 201, 206
  type, 92, 103, 194, 196
  velocity, 95, 208
Electrolyzing current, 78, 84
Electron beam machining, 9, 157, 160, 179
Electropolishing, 70–75
Electrostream drilling, 106, 108
Environmental impacts, 46, 99
Equilibrium gap, 86, 87
Erosion phase, 206, 222, 223
Etch factor, 56–58, 60, 61
Etch rate, 58, 61–63, 174
Etchant, 10, 12, 57–60, 63, 66, 67, 70
Evaporation, 118, 143, 152, 154, 158, 203, 218, 223

Faraday's laws, 77, 85, 88, 101, 110, 184
Feed rate, 35–37, 40, 45, 46, 80, 83, 84, 86, 87, 90, 93, 95–97, 100, 101–105, 112, 130, 171, 186–188, 208–210, 223, 224, 243
Fiber-reinforced plastics, 34, 36, 37, 155
Flushing, 12, 22, 115, 118, 120, 121, 123–125, 178, 182, 203, 209, 214, 219, 220, 226
Fused deposition modeling, 235, 237, 244

Gap:
  pressure, 96, 101, 208
  short circuit, 87, 208, 220

voltage, 81, 83, 85, 90, 95, 96, 102, 106, 113, 119, 137, 183, 184, 189, 194, 208, 219, 220, 224
width, 90, 92, 93, 115, 119, 183, 186, 205
Graphite, 23, 27, 30, 31, 36, 47, 95, 121–123, 126, 128, 135, 152, 211, 213, 214, 218, 228
Grinding:
  action, 187, 221
  process, 184
  stick, 194
  wheels, 7, 152, 179, 186
Grooving, 149, 151, 177, 179

Heat-affected zone, 9, 35, 70, 118, 129, 150, 162, 171, 213
History of machining, 1
Holographic interference solidification, 232
Hybrid electrochemical processes, 182
Hybrid thermal machining, 11, 12, 203, 204
Hydraulic pump, 32
Hydrogen, 59, 64, 79, 99, 101, 168

Ice jet machining, 9, 46, 52
Ignition delay, 117, 210, 220
Insulating ceramics, 133, 134, 179
Integrated circuit, 65, 163, 164
Intensifier, 32, 33, 44
Ion beam machining, 3, 4, 9, 172, 174
Iron, 1, 49, 78, 79, 82, 95, 161

Jet:
  cutting nozzle, 33
  diameter, 34, 41, 46, 108
  machining, 9, 32, 34, 35, 39, 41, 43, 46, 52, 145
  velocity, 33, 34, 41

Laminated object modeling, 245
Laser-assisted electrochemical machining, 11, 181, 199
Laser beam:
  drilling, 147
  machining, 9, 140, 144, 146
  texturing, 151
Liquid-based techniques, 230
Liquid thermal polymerization, 235

Machinability, 4, 6, 23, 75, 83, 89–92, 104, 113, 143, 144, 160–162, 169, 170, 174, 177, 178, 186, 187, 201, 210, 211, 223, 226

Machining:
  by abrasion, 6
  by cutting, 5, 6, 13, 52, 113, 178, 201
  of fiber-reinforced plastics, 36
  of spheres, 131
  system, 15, 16, 32, 39, 44, 46, 48, 50, 102, 120, 135, 166, 167, 172, 183, 189, 190, 198, 204, 205, 215, 218
Magnetic finishing of:
  balls, 50
  magnetic tubes, 50
  nonmagnetic tubes, 51
  rollers, 52
Magnetostrictor, 15–21
MA phase, 185, 222, 225
Maskant, 55, 57–59, 61
Mechanical abrasion, 5, 9–12, 22, 39, 50, 181, 184–186, 189, 190, 193, 194, 197–199, 203, 214, 221–223
Mechanical amplifier, 15, 20
Mechanical machining, 9, 10, 12, 181, 203, 224, 226
Mechanical pulse electrodischarge machining, 220, 227
Mechanism of material removal, 115, 172, 174
Metallic bond, 183, 185, 189, 192, 193, 214, 215
Metal matrix composites, 149, 178
Micro-ECM, 98, 137, 155, 177
Micro-EDM, 135
Micromachining, 4, 13, 74, 113, 122, 130, 155, 157, 178, 201, 246
Micro ultrasonic machining, 31, 52
Multijet modeling, 238

Nickel, 21, 58, 61, 62–64, 67, 73, 82, 88, 89, 91, 144, 162, 169, 176 187
Nitric acid, 103
Nontraditional machining, 5, 8, 9, 11, 229
Numerical control, 3, 29, 31, 132, 171

Overcut, 26, 80, 83, 100, 101, 103, 108, 130, 186, 187, 214
Oxide film, 89, 90, 102, 185, 193–195, 197
Oxygen, 39, 71, 79, 128, 148–150, 168, 169, 178, 210, 226, 243

Passivation, 12, 85, 90–92, 95, 96, 181, 198, 201, 223
Photochemical milling, 59, 66
Photoresist, 59, 66–68, 175

Plasma:
  arc, 166, 167, 169, 171, 172
  beam, 9, 166, 169, 170
  channel, 118, 119, 135, 216, 218, 220
  jet, 166–169
Polishing, 4, 5, 7, 30, 31, 38, 39, 42, 48, 50–52, 70–75, 91, 109, 128, 172, 176, 196, 218
Powder-based processes, 241
Printed circuit boards, 34, 38, 52
Process:
  accuracy, 26, 90, 95, 101, 195
  capabilities, 45, 104, 106, 158
  characteristics, 41, 42, 87, 189
  control, 95, 137
  description, 46, 66
  parameters, 34, 61, 73, 103, 113, 114, 165, 184, 198
Pulse:
  charge, 161, 162
  current, 97, 98, 116, 120, 126–128, 134, 135, 206
  duration, 127, 138, 142, 146–148, 154, 160, 161, 163, 209
  electrochemical machining, 113, 208
  energy, 126–128, 147, 148, 161, 162, 210, 223
  frequency, 160, 165, 209, 214
  on-time, 120, 127
Pulsed power supply, 221

Rapid prototyping, 229, 232, 235, 240, 246
Recast layer, 35, 65, 104, 118, 129, 131, 146, 166, 204
Resolidified layer, 129, 152
Rotary ultrasonic machining, 28, 52

Sapphire, 28, 33, 34, 41, 42
Sawing, 5, 130, 188
Scribing template, 57, 60
Selective laser sintering, 230, 241, 243
Servo controlled feed, 121, 213, 215
Shape deposition manufacturing, 240
Shaped tube electrolytic machining, 102
Shielded plasma, 167–170
Short circuit, 80, 81, 87, 90, 119, 208, 209, 218, 220
Shot blasting, 134, 151
Slotting, 159, 160, 163
Slurry injection methods, 22
Smoothing, 12, 66, 71, 74, 75, 175, 176, 197

Sodium chloride, 78, 81, 82, 189
Sodium nitrate, 81, 82, 189
Solid-based processes, 230
Solid foil polymerization, 244
Solid free-form fabrication, 229, 240
Solid ground curing, 233
Spark:
  erosion, 132, 226
  gap, 135, 212
  machining, 178, 179, 226
Standoff distance, 34, 37, 40, 41, 45
Steels, 1, 35, 61, 62, 95, 106, 111, 168, 169, 186, 187, 218, 226
Stereolithography, 230, 231, 235, 242
Surface:
  effects, 175
  formation, 93, 208
  integrity, 4, 127, 186, 226
  quality, 1, 2, 8–10, 12, 24, 26, 27, 35, 37, 46, 50, 62, 64, 73, 83, 92, 93, 95, 99, 101, 123, 125, 160, 162, 169, 181, 187, 188, 197, 199, 206, 209
  reflectivity, 147, 165
  roughness, 27, 30, 36, 41, 42, 45, 62, 63, 71, 73, 93–95, 111, 126–128, 131, 135, 162, 165, 177, 190, 195, 206, 209, 210, 213, 218, 220, 221, 223
  treatment, 38, 43, 177

Texturing, 38, 53, 128, 130, 134, 135, 151, 175, 177, 178, 179
Theory of ECM, 78
Thermal energy method, 109
Thermal properties, 125, 143, 146, 160, 161
Thermal machining, 9, 11, 12, 203, 204, 227
Thinning of parts, 65
Three-dimensional printing, 243, 244, 246
Titanium, 23, 39, 58, 59, 61–65, 67, 82, 84, 90, 91, 103, 106, 107, 143, 144, 161, 187, 188, 218, 243
Tool:
  design, 83, 96, 113, 138
  electrode, 97, 99, 103, 110, 113, 115, 119, 121
  feed rate, 83, 95–97, 102, 103
  insulation, 84, 90, 101, 111
  material, 1, 23, 27, 31, 121, 122, 208
  oscillation, 24, 27, 192, 198, 206
  shape, 15, 25, 83, 97
  wear, 6, 13, 27, 74, 118, 119, 122, 136, 155, 156, 177, 184, 199, 220, 224

Index

Tooling for CHM, 57
Traditional machining, 5, 8, 35, 67,138, 218
Traditional grinding, 187
Trimming of electronic components, 154
Tungsten, 21, 23, 26, 39, 42, 82, 84, 91, 113, 121, 123, 136, 157, 161, 162, 166, 168, 172, 187, 188, 216, 218
Turbine blade, 29, 104, 106, 130, 163
Turning, 3, 5–7, 83, 99, 112, 138, 139, 151, 167, 169–171

Ultrasonic accuracy, 26
Ultrasonic-assisted electrochemical machining, 12, 181, 197
Ultrasonic contour machining, 53
Ultrasonic machining, 9, 12, 15, 28, 31, 52, 145, 197, 218
Ultrasonic machining system, 16
Ultrasonic polishing, 30, 31
Ultrasonic sinking, 29

Ultrasonic vibration, 22, 28, 124, 203, 218–220, 227
Vaporization, 109, 134, 141, 142
Volumetric removal rate, 41, 88, 126, 142, 159, 184, 208, 210

Wire:
  cutting, 132, 140, 204, 205, 209, 226
  EDM, 43, 104, 106, 115, 132, 133, 136, 140, 178
  electrodischarge grinding, 136
  stripping, 38
Workpiece:
  impact hardness, 25
  material, 4, 6, 9, 12, 15, 23, 26, 32, 38, 39, 43, 45, 49, 55, 58, 61, 63, 73, 81, 82, 99, 121, 125, 126, 129, 135, 143, 145, 151, 159, 160, 161, 169, 170, 175, 179, 184, 196, 203, 213, 223
  shape, 83, 122

ABOUT THE AUTHOR

Hassan El-Hofy is a professor in the Mechanical Engineering Department of Qatar University and an associate editor for *Qatar Engineering Journal*. He is also the chairman of the College Accreditation Committee (ABET) for the Mechanical Engineering Department and has 50 research papers related to nonconventional as well as hybrid machining processes.

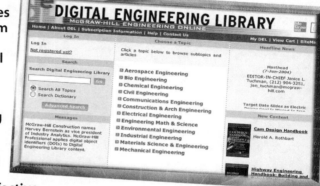